I0481413

ISBN-13: 978-1981884162

DARK MATTER and NEUTRINOS: Solved?

3 New Theories

Dark Matter

The theory presented herein will address what is considered to be one of the most important and elusive subjects in astrophysics of the 21st century; namely, "What Is Dark Matter?"

My goal is to set forth herein a theory as to the CAUSE of Pressure Ether/Dark Matter (my terminology).

Neutrinos

The writing presented herein consists of defining and exploring a very popular 21st century subject; namely, the Neutrino. It was first described as the "ghost particle" due to its nature of being totally unable to be detected. As of today, it remains as such.

My goal is to set forth a substantial amount of known facts at the present time and to then evaluate the results. Two new theories are listed as solutions for the Neutrino problem.

CONTENTS

For those who think differently.

PREFACE

When it comes to stepping outside the norm, this is when the word "theory" comes into play. The idea of thinking of a concept so new that it even astounds you, the creator of the theory, this moment can leave one totally speechless at times.

"Why am I having this natural flow of a never-before-thought-of-point-of-view?" The logic is there, the connection to the objects in question have been thoroughly researched, and the automatic writing of the theory which comes so naturally are all parts of the amazing process that develops into a new discovery.

As for the word "discovery" even as it relates to science, this exact accomplishment has been stated very eloquently in years past as follows:

(1851) Arthur Schopenhauer, German philosopher: The task is, not so much to see what no one has seen yet; but to think what nobody has thought yet, about that which everybody sees.

(1957) Albert Szent-Gyorgyi, Hungarian biochemist: "Discovery consists of seeing what everybody has seen and thinking what nobody has thought." (as per Schopenhauer)

It's as simple as that.

Therefore, the discovery of a new scientific theory can barely be explained at times. Its actual happening may have been instantaneous, but it most assuredly will have taken years to piece together. Nevertheless, when the complete theory has been painstakingly brought to pen and paper for all to see with an assurance of bursting confidence, it is the duty of the creator to share its findings. Thus, a new theory has been born.

The following one sentence below makes so much sense, written by the beloved intellectual humorist Mark Twain, aka Samuel Longhorne Clemens (1835-1910). It appears in a book "LETTERS FROM THE EARTH" (written approx. 1905; published posthumously 1962), as he was stating the situation whereby "Adam" (as in Adam and Eve, the first beings of the Earth) is particularly engaged in researching, studying, and learning as a "scientist" all things that he came across in nature while living in the Garden of Eden. His first great curiosity, which could be called his first scientific discovery, was that *water ran downhill, and not up*. Can you just imagine what his delight must have been after so many, many, many days of observation and experimentation with the action of water always flowing downhill, but never up? It had to be sheer exhausting exhilaration.

Therefore, Twain states so beautifully: **"Knowledge has to be acquired by hard work; none of it is flung at our heads gratis."**

* * *

INTRODUCTION

The theory presented herein will address what is considered to be one of the most important and elusive subjects in astrophysics of the 21st century; namely, "What Is Dark Matter?" That was the opening sentence to my writing "THE THEORY OF THE <u>CAUSE</u> OF DARK MATTER/PRESSURE ETHER, One-Half of the Puzzle Solved, March 2010."

By way of explanation, I spoke of choosing the subject of LIGHT to occupy my time upon moving to my hometown in Pennsylvania in November 1999 to recuperate from a second failed back surgery of 1998. Being disabled in walking at the young age of 60 years old but having a small Public Library right across the street, I was able to embark on a total learning program into this chosen subject of LIGHT with all of its many attributes and functions. In the ensuing years, that subject would lead to a profound interest in the workings of the Sun and into the subject of Dark Matter, which although known to be in vogue at the time was entirely outside my knowledge.

However, what I found so terribly strange in all of my research and reading was the fact that scientists all were completely focused on **capturing** Dark Matter. There existed many different experiments by which they were assuming that Dark Matter consists of *particles* that could be captured; even though Dark Matter was constantly stated as being invisible and of an unknown substance!

Please note that my terminology for this substance is "Pressure Ether" and will be used interchangeably within this text. It should also be noted that this is only one-half of the puzzle solved. The actual COMPOSITION of Dark Matter can only be hypothetically determined, hopefully, once the cause can be evaluated. This I leave to the experts. My goal is to set forth herein a theory as to the CAUSE of Pressure Ether/Dark Matter.

UNIVERSAL PURPOSE OF DARK MATTER

In the distant past, and before the vision and institution of the concept of Dark Matter existing in the cosmos, it would be safe to say that the common notion was that Gravity must be the key system which is holding everything in its place in the Universe. But this cannot be. It became my personal belief that in order for Gravity to work or perform, there has to exist two opposing entities which mutually attract each other core-to-core (the inner core of each body), or in the case of small objects such as asteroids, this would be due to their overall MASS. According to Isaac Newton's thinking, the journey of an object in the confines of Gravity will be determined by the object's MASS and its speed of travel; i.e., MASS will pull on MASS. It's as simple as that.

In any case, there seems to be a basic law that cores hold other cores in a locked obedience, depending upon the volume of each MASS passing each other and its speed of travel. Hence, the inner base of a large planetary object, such as Suns and Planets, could be the main mechanism for grabbing another planetary object passing within its reach.

This is all well and good for holding closely aligned (in universal terms) objects in Space in a pattern of stability. However, there are vast, vast areas within Space where planetary objects do not exist. The question begs: What is holding this emptiness of Space "in a pattern of stability"? In these confines, Gravity has no function. It is totally amiss. Further, if a galaxy island of stars are all held together by their actions of gravitational pull upon one another, then what is holding the *entire island of stars* "in a pattern of stability"?

For the 21st century astrophysicist and astronomer, the answer is now Dark Matter.

PERCENTAGE OF DARK MATTER IN THE UNIVERSE

As stated in my 2010 document Preface: "The subject of <u>Dark Matter in the Universe</u> is being taken quite seriously at the present time with mathematical calculations of 4% Atoms, 21% Dark Matter, and 75% (brand new) Dark Energy. However, the foremost question as to the **genesis** of this unknown substance has not gained any substantial ground, to my knowledge." To be sure, there is nothing definitive about any of these figures, as they have all changed in the present time. They are, of course, all hypothetical valuations.

However, my only dismay at the above figures – in 2010 and in the present 2017 – is the small percentage given to *quote* "Dark Matter" *end quote* and the very large figure given to *quote* "Dark Energy" *end quote*. Suffice it to say that *without knowing precisely what the composition would be of Dark Matter, then how can such a figure be determined for something brand new called "Dark Energy" in the Universe?*

Personally, I have no belief that there is a separate Dark Energy component. Since all matter can be said to contain energy of a reinstituted state, and some matter can exhibit kinetic energy, it seems logical that Dark Matter/Pressure Ether in and of itself may prove to have some kind of energy value as it relates to its surroundings. Nevertheless, to presume that there is a separate action for Dark Matter material as opposed to a possessed 75% Dark Energy action whose purpose has yet to be stated (other than the usual expansion of the Universe statement) is, in my opinion, erroneous.

At this point, it might be helpful to realize that: Gravity is a force; not a substance. Electricity is a force; not a substance. Pressure Ether/Dark Matter is a substance; not a force.

Finally, it should be remembered that although Gravity may be reported to be an **exerting force** being manifested between objects, Dark Matter would primarily be the overwhelming **binding substance** that holds objects together in their place in the Universe.

AETHER VERSUS DARK MATTER

Before the words "Dark Matter" were created to envision whatever is holding everything in the Universe in its place, historically the word "AETHER" (or "Ether") was first used by the Ancient Greeks. The acceptance of this substance (for lack of a better word) was a very natural position. Of course, its composition was not known nor is it written anywhere, to my knowledge, that it was generally pursued at that time. From a historical perspective, even though the Ancients did not know of other galaxies or even the Milky Way Galaxy as a whole, these learned men thought about our Solar System and knew there had to be an element which was keeping everything in an orderly fashion.

Aristotle (384-322 B.C.) taught in 350 B.C., that the "celestial spheres which surround the Earth consist of a single incorruptible element, AETHER." Although there are no spheres as such, it was believed that these celestial spheres supported the Wandering Stars, the name for Planets. This amazing thought-experiment is that these spheres were said to be filled with a single incorruptible element: AETHER. There was no doubt that there was a substance in the heavens which held everything in its place.

Throughout the annals of Time and writings, the word "Ether" has been resurrected over and over again. But its unverifiable status has always left it skeptical as well as shunted. When the terminology was not needed to validated or be integrated into any of the subject matters being conceptualized, the Ether was time and time again dismissed, dropped, or eliminated as unnecessary for the propagation of a particular aspect of theory. Because of this, a serious void was created in the knowledge of Empty Space for quite some time. But now it's called DARK MATTER. Today the quest is on in the 21st century to solve the existence and verify the validity of the Ether now known under a different name.

As stated in my 2010 document, the quest is in earnest today to find the substance that holds the Universe together. Whatever name is chosen for this substance, such as AEther - Ether - Plenum - Dark Matter - Pressure Ether - Quantum Foam - Plasma - Quintessence, the old adage holds true that a Rose by any other name is still a Rose.

PROCESS OF ELIMINATION FOR THE CAUSE

Since Dark Matter aka Pressure Ether is deemed to inhibit every nook and cranny of the atmosphere of the Universe, then it stands to reason that something which exists in great quantity would be the culprit for causing the substance to appear and/or be created in a non-stop fashion throughout universal Space. In this context, the **continuous production** and build-up of Dark Matter cannot be overemphasized. The search for planetary objects which could contribute to cause this Pressure Ether to come into existence is rather small; i.e., there are only **5 primary universal objects so prevalent in the cosmos.**

Therefore, these objects can be looked to for a linkage to the creation of Dark Matter/Pressure Ether, as summarized below.

1. Suns: Due to the fact that stars dominate the Universe, a logical conclusion can be deduced that suns are somehow the bearers of a causative condition for the creation of the Ether.

2. Planets: Because planets may not exist in a star's evolutionary "solar system" processes, if any, these rotating soil, ice, or gas encrusted bodies can be eliminated.

3. Moons: Wherefore moons are dependent upon adjacent planets for existence, their obvious slim numbers make them ineligible.

4. Asteroids (Planetary Rings): It is my firm belief that asteroid rings encircling a planet are merely "failed moons" and here again their lack in numbers would make them a non-contributor.

5. Comets (Meteors): Comets as traveling objects in the Universe indicate that they could possible excite an energy chain reaction along their lengthy journeys. However, they are simply not in an over abundant state within an ever-expanding Universe. Meteors are also in short supply, and they simply dissipate or dissolve within their travels. Neither of these objects would be a serious contender.

This simple process of elimination strongly suggests that Suns per se must be substantially contributing to the cause of Dark Matter/Pressure Ether in some yet to be discovered way.

DARK MATTER DEFINITIONS

Please know that there are a host of definitions for Dark Matter and its hypothetical percentage of occupancy in the Universe. The following 13 sources are examples of the most prevalent definitions.

Encyclopedia Britannica

Dark Matter: a component of the universe whose presence is discerned from its gravitational attraction rather than its luminosity. Dark Matter makes up 26.5% of the matter-energy composition of the universe; the rest is dark energy (73 percent) and "ordinary" visible matter (0.5 percent).

Collins English Dictionary

Dark Matter: matter known to make up perhaps 90% of the mass of the universe, but not detectable by its absorption or emission of electromagnetic radiation.

The Free Dictionary

Dark Matter: material that is believed to make up nearly 27% of the mass of the universe but is not readily visible because it neither emits nor reflects electromagnetic radiation, such as light or radio signals. Its existence would explain gravitational anomalies seen in the motion and distribution of galaxies. Dark Matter can be detected only indirectly, e.g., through the bending of light rays from distance stars by its gravity.

Science Dictionary

Dark Matter: material in the universe, invisible to electromagnetic radiation, whose presence is implied by its gravitational effects. It probably consists of certain elementary particles, undetectable dim stars or planets.

NASA Science Newsletter

Dark Matter: By fitting a theoretical model of the composition of the universe to the combined set of cosmological observations, scientists have come up with the composition of 68% dark energy, 27% dark matter, 5% normal matter.

National Geographic

Dark Matter: ordinary, or baryonic, matter makes up less than 5 percent of the mass of the universe. The rest of the universe appears to be made up of a mysterious, invisible substance called dark matter (25 percent) and a force that repels gravity known as dark energy (70 percent).

CERN

Dark Matter: Researchers have been able to infer the existence of dark matter only from the gravitational effect it seems to have on visible matter; making up about 27% of the universe. The matter we know only accounts for 5% of the content of the universe. Dark energy makes up approximately 68% of the universe and appears to be associated with the vacuum in space.

Wikipedia

Dark Matter: Fully 70% of the matter density in the universe appears to be in the form of dark energy. Twenty-six percent is dark matter. Only 4% is ordinary matter.

Merriam-Webster

Dark Matter: nonluminous matter not yet directly detected by astronomers that is hypothesized to exist to account for various observed gravitational effects.

MacMillan Dictionary

Dark Matter: a substance that scientists think exists out in space, but for which they have no direct proof.

Thesaurus

Dark Matter: a hypothetical form of matter that is believed to make up 90% of the universe; it is invisible (does not absorb or emit light) and does not collide with atomic particles but exerts gravitational force.

Extreme Tech

Dark Matter: By modern estimates, the universe is only about 5% regular matter and energy, and about 27% dark matter, or more than five times as much. The remaining 68% of the universe is thought to be dark energy.

Ask An Astronomer (2015)

Dark Matter: We believe the cosmos to be composed of roughly 0.03% heavy elements (anything other than hydrogen and helium), 0.3% neutrinos, 0.5% stars, 4% free hydrogen and helium, 25% dark matter, and 70% dark energy.

NOTE: For anyone wanting to know the early history of the inference/theory discovery of Dark Matter, the following few names are given here: Lord Kelvin (1884); Henri Poincare (1906); Jacobus Kapteyn (1922); Jan Ort (1932); Fritz Zwicky (1933); and Vera Rubin and Kent Ford (1960s/1970s).

EXPERIMENTS TO "DETECT" DARK MATTER

The following six experiments are currently in operation to detect Dark Matter.

1) AMS Alpha Magnetic Spectrometer, a sensitive particle detector on the International Space Station, installed in 2011. Statement made: "We have measured an excess of positrons [the antimatter counterpart of an electron], and the excess can come from Dark Matter. But at this moment, we still need more data to make sure it is from Dark Matter and not from some strange astrophysics sources. That will require us to run a few more years." It is said that AMS has tracked more than 100 billion cosmic ray hits in its detectors.

2) Beneath a mountain in Italy is the LNGS's XENONIT which is hunting for signs of interaction after WIMPS [Weakly Interacting Massive Particles] collide with xenon atoms; (with ultra-low background massive detectors on Earth).

3) LUX Large Underground Xenon dark-matter experiment, in a gold mine in South Dakota.

4) IceCube Neutrino Observatory, an experiment buried under Antarctica's ice is hunting for "sterile neutrinos." Sterile neutrinos are said to only interact with regular matter through gravity, making it a strong candidate for dark matter.

5) The European Space Agency's Planck spacecraft has been building a map of the Universe since launched in 2009. By observing how the mass of the Universe interacts, it can investigate both dark matter and its co-partner, dark energy.

6) NASA's Fermi Gamma-ray Space Telescope (2014) made maps of the heart of the Milky Way in gamma-ray light, which revealed an excess of emissions extending from its core. Statement made: "The signals we find cannot be explained by currently proposed alternatives and is in close agreement with the prediction of very simple dark matter models." Statement made: "When two dark matter particles crash into each other, they might release a gamma ray."

Needless to say, the hunt for Dark Matter in the form of particles is certainly ongoing. In doing so, each one assumes that this invisible substance has the properties and/or components of a particle; albeit an unknown or exotic particle to be discovered.

DWYER'S DISAGREEMENT WITH GALAXY CLUSTERS

Before presenting the theory of solving the CAUSE of Dark Matter, one more important phase of the current belief of what happens within galaxy clusters in the Universe needs to be addressed. It has been written that galaxy clusters are the largest gravitationally bound objects in the Universe. Scientists have made a list of 3 major components to a galaxy cluster as follows:

1. A galaxy cluster consists of hundreds of galaxies obviously containing stars, hot gas, and dust. However, what causes these galaxies to group together in a cluster form is not known.

2. A galaxy cluster has clouds of hot gas, designated at 30-100 million degrees Celsius in temperature, existing within the space between galaxies. This gas is said to be invisible to telescopes.

3. Dark Matter that has not been able to be detected by any type of telescope is stated as making up a significant amount of the galaxy cluster because its "presence is felt" through its "gravitational pull" on the galaxies and the hot gas. It is believed that although the galaxies and hot gas are tremendously massive, scientists have deduced that it would take something with 10 times more MASS needed in order to hold a galaxy

cluster together, and that something has been deduced to be the unknown Dark Matter. Consequently, it is stated that Dark Matter must exist to provide the additional gravity which is holding any galaxy cluster together.

This last part of explaining what holds a galaxy cluster together is, in my opinion, terribly <u>wrong</u>. Scientists have taken galaxies and the gas between galaxies, and then proposed that something is missing in a gravitational sense for holding these galaxies together. Therefore, by theory that the missing something is Dark Matter, they go a step further to <u>pronounce</u> that this invisible Dark Matter **has gravity as a component**!

My sequence of events as to galaxy clusters is as follows:

Each individual galaxy would naturally contain Suns, Planets, and Moons in their separate development. Gravity would definitely exist as a core-to-core pull on these objects. That is, Suns keep their Planets gravitationally obedient, and Planets keep their Moons gravitationally obedient. That is the main function of Gravity. [Note: Suns will also keep their binary Suns in obedience.)

Hot gases should exist between each individual galaxy cluster by a natural occurrence in varying temperature conditions as the rotation of these galaxies have their <u>weather pattern developing</u>. However, these clouds of gas contain NO GRAVITY to exert on any large planetary object. They are merely gas build-ups in the region.

Scientists are still unanimous that the actual development of a cluster of galaxies remains an "unknown" item. Why would a single individual galaxy be pulled into a close contact (in universal terms) with another galaxy? The answer perhaps might be the pull of gravity that exists at the <u>edges</u> of a galaxy. (It was Vera Rubin in the 1970s who measured stars in several galaxies, including Andromeda, and showed that stars at the edge of a galaxy have the same rotation velocity as stars near the center of a galaxy; meaning, of course, that stars at the edge of the galaxy do not move slower, and that "something" within the galaxy was keeping everything in its place.) An individual galaxy joining up into clusters or groupings would seem to be the result of a close encounter with another individual galaxy in order for Gravity to exert any force. These galaxies do not collide; they merely attach to the group surroundings.

However, the answer to the above question is *not gravity* being exerted at the edges of any galaxy. The answer, in my way of envisioning events, is that the compactness of Dark Matter around each individual galaxy has <u>pushed</u> these formations into a grouping situation. By sheer <u>density</u> of Dark Matter, which exists as an "Ether Web" (my terminology), this substance has slowly pushed those galaxies into what is termed a cluster grouping. At this point, it is my firm belief that **Dark Matter does not exert Gravity**. It must be realized that galaxy clusters will form very slowly in a natural process of Time.

Further, Dark Matter as "the glue that holds everything in its place in the Universe" (my original terminology) naturally abounds within each individual galaxy object, as well as outside of any formed galaxy. This Ether Web or cocoon-type structure (my terminology 2009) permits huge planetary bodies to pass through it. Pressure Ether/Dark Matter has no need to exert any aggressive force of disturbing energy upon planetary bodies. My statement in 2012: **"Its sole purpose is to remain and be non-hindering as well as non-binding and to hold everything in its place in universal Space."**

In other words, all movements of objects in the Universe from the incomprehensible greatest to the absolute quantum smallest will take place on a continuous basis – YET – be held in place in the scheme of things by a web of substance, a buoyancy blanket of substance, which Man has decided to call Dark Matter. It can be envisioned that Pressure Ether aka Dark Matter would not exert any *force* upon planetary objects due to its composition (as yet unknown) unless the density of the surrounding MASS would greatly outweigh the object, thus pushing or nudging the object into a further direction. But in any event this action is *non-hindering* to the planetary object itself and only expresses the need to expand the space between objects.

Further it is imperative that the composition of Pressure Ether does not overpower any object to the point of then drastically and haphazardly rearranging the heavenly bodies. My conclusion is that the purpose of Dark Matter is not to be a force for expansion but to be a substance which holds everything in its place; i.e., a buoyancy blanket for all universal objects to **reside**.

Finally, it is assumed that the force of Gravity is the cause of an object maintaining its orbit in Space. This is true as can be seen for the close proximity of the Planets and the Sun in our magnificent Solar System or any other Solar System in the Universe. However, where singular Suns exist with no Solar System to their credit, these Suns must therefore reside alone and with no attractions to any other planetary body. Hence, Gravity as we know it has NO PURPOSE in this situation. The only active ingredient would be Pressure Ether/Dark Matter to hold these singular Suns in their present place within the Ether Web. It must be remembered that Dark Matter per se permeates the entire Universe completely.

Of course, those singular Suns can move freely as the eons pass, simply by traversing *through* the bulk of Dark Matter present, which to my way of envisioning this substance is non-hindering and non-binding in its manner. Unfortunately, by science giving Dark Matter the component of Gravity and equating it to be an **energy** being exerted from and/or because of a bulk of this substance, a new category with its hypothetically overwhelming percentage valuation has been theoretically created! A ratio was stated in my 2010 document to be that mathematical calculations of the Universe had been decided at that time to be: 4% Atoms, 21% Dark Matter, and 75% (brand new) **Dark Energy**. As anyone can see, this over-extended Dark Energy ratio for a percentage in the Universe regarding an action of energy proliferation is totally out of balance. In my humble opinion, the term Dark Energy should be abolished.

Needless to say, if Dark Matter would exert a "movement" of itself in some such fashion due to its density build-up, it certainly has no need to exert any aggressive force of "disturbing energy" upon any planetary bodies in the Universe. Its sole purpose is to be a **buoyancy blanket** for all universal objects to reside within. The scientific community should come to the realization that if a Dark Energy function is to be listed as a percentage in the Universe, then its purpose, as well as its genesis, would definitely have to be acknowledged outside of the realm of Dark Matter. The two are not connected.

ISAAC NEWTON'S OPTICKS

When my two writings of 2008 and March 2010 were completed, the book OPTICKS by the illustrious Sir Isaac Newton (1704) was purchased and subsequently received on July 8, 2010. Here again, this was done to further my studies re LIGHT. To begin with, Newton's eight definitions pertaining to LIGHT are absolutely stunning in scope. By reading many parts of the book verbatim but skipping many pages of descriptive experiments, I quickly became drawn to page 338 where he embarks upon proposing a set of Queries regarding the many aspects of LIGHT. My interest became intense with Query 18 wherein he begins serious questions regarding the Medium. (This word "Medium" could be interpreted, of course, as Dark Matter in today's terminology, but it is spoken of as AEther, which is the language of his day.)

Therefore, on July 14, 2010, it was incredible to read in Query 21: *quote* "And that the elastic force of this Medium is exceedingly great, may be gather'd from the swiftness of its vibrations." *end quote*

That is to say, Newton envisions that AEther may be the product of **fast vibrations**, and further he states very clearly that the vibrations must be swifter than LIGHT and obviously swifter than Sound. And although he is speaking of LIGHT moving from the Sun, he never explicably claims that this is the **cause** of the AEther. But he speaks of the vibrations of the AEther as the cause of the AEther.

In making a reference to Particles, then Newton states quite clearly and unashamedly about the AEther in Query 21: *quote* "And so if any one should suppose that AEther (like our Air) may contain Particles which endeavor to recede from one another (for I do not know what this AEther is) and that its Particles are exceedingly smaller than those of Air, or even than those of Light: The exceeding smallness of its Particles may contribute to the greatness of the force by which those Particles may recede from one another, and thereby make that Medium exceedingly more rare and elastick than Air, and by consequence exceedingly more able to press upon gross Bodies, by endeavouring to expand itself." *end quote*

In my March 2010 theory regarding the cause of Dark Matter/Pressure Ether, my pictorial mind-experiment was to envision that LIGHT is the Agent that constitutes the production of the Ether, which is facilitated by the Friction invoked by the Speed of LIGHT through Empty Space. Then especially later after reading the Query section of Newton's book <u>OPTICKS</u> of 1704, I came away with the thought that he was perhaps *so close* to claiming that LIGHT per se was the *cause* of an AEther, but instead chose to have an unknown-caused AEther **self-replicate via vibrations** and thereafter perpetuate the constant flow of same. This self-replication of the AEther is very similar to one of the mistaken concepts that I entertained in 2005 regarding Amoeba Particles via Solar Wind, which was abandoned due to the cessation of the flow of Solar Wind. Nevertheless, the fact that Sir Isaac Newton's presentation of 1704 and my very first pictorial mind-experiment of 2005 were on the same path is purely coincidental.

<u>THE SUBJECT OF LIGHT</u>

There is an **order** in the Universe that may be beyond human comprehension in the 21st century. However, Man is confident that Suns are the dominant feature of the cosmos and have even designated categories, such as: Red Dwarf, White Dwarf, Brown Dwarf, Main Sequence Star, Neutron Star, Variable Stars, Red Giant, and finally Supergiants and Hypergiants. Nevertheless, all of these Suns are born in the same identical manner (which has yet to be agreed upon), and their eventual product is the illumination of LIGHT. Man in the 21st century is still trying to understand the subject of LIGHT with its many,

many facets. Here is a list of only a very few well-known terms that one will encounter in their quest to understand the subject of LIGHT: redshift, wavelength, tired light theory, scattering, cosmic microwave background (CMB), thermal radiation, absolute zero, and light quanta.

In reading the July 2012 issue of Scientific American, there was a section entitled "Nobel Pursuits." Because of the year's 62nd annual meeting in 2012 of scientists and veteran Nobel Prize winners focusing on physics, the magazine listed a few vignettes of past winners featured in their magazine on various topics. The one article of complete interest to me was entitled "What is Light?" (published in April 1928 by Ernest O. Lawrence and J. W. Beams; Nobel Prize in Physics in 1939 by Lawrence). The writing states that the question of what exactly is **"a quanta"** in terms of light length was actually committed to experiment, and an answer was definitely obtained. It was determined that the length of **"a light quanta"** is less than a few feet long. Whether or not further work has been done on this subject is something that might prove useful to pursue in the scientific community future because it would relate to the Speed of Light which I have suspicions may be a variable.

Then in connection with the subject of LIGHT, we now have the subject of neutrinos. Very briefly, it seems that in September 2011, neutrinos moving faster than the Speed of Light were detected as they traveled from CERN laboratory to the Gran Sasso laboratory. Subsequently, four additional experiments were done. Then on June 8th at the International Conference on

Neutrino Physics and Astrophysics in Kyoto, it was announced that all four experiments had shown a neutrino time of flight was not faster than the Speed of Light but was in fact below that figure. The previous test was ascribed to basically faulty equipment. And currently, a few physicists have already speculated that neutrinos are simply another form of LIGHT in such a minute composition and possessing, of course, the dynamic Speed of Light (or even more?). Needless to say, this subject with its connection to the subject of LIGHT is currently being examined quite intensely in the 21st century.

COMPOSITION OF PHOTONS/LIGHT

Photons equal LIGHT; i.e., they are one and the same, using two different names. Albert Einstein is responsible for the popularity of the word "photon" in the early 1900s as he proposed the existence of energy packets "during the transmission of Light." However, it was Max Planck who previously explained in the very early 1900s that heat radiation (Light) is **emitted and absorbed** in distinct units called "quanta." Therefore, the energy of a photon depends on the radiation (Light) frequency being dispensed in quanta packets, and Light (photons), of course, travel within a vacuum at the Speed of Light, which is 186,172 miles per second.

In my humble opinion, the word "photon" is just a capsulized version of talking about Light as packets of energy. That is, electromagnetic radiation is the flow of photons through Empty

Space. Light aka photons are bundles of energy which are always moving at the Speed of Light as they behave like a particle and as a wave; the particle-wave duality theory was finally decided upon by scientists that Light (photons) will have properties in wave mechanics, but they are also treated as particles. Simply stated: photons are the fundamental tiny little particles of Light.

It is important to always remember that a photon (a particle of Light) cannot decay on its own as it has NO MASS. However, it has been decided that Light (as a photon particle) has ENERGY which can underline interact with other particles re energy transfer or creation. This is the reason that a photon has been given a wave category as well as a particle category. It all depends on the situation encountered (of which many can be entertained). As everyone knows, LIGHT still remains an unknown subject regarding its propagation and its encounters. NOTE: My one main question is: Why would a photon be considered "massless" if it can contain energy and motion?

SEARCHING FOR THE CAUSE

In searching for a catalyst which would create an unknown substance throughout the Universe, it was imperative that there be a *continuous happening* in the Void of Empty Space that would be capable of being the cause of generating a never-ending substance. The use of my terminology "the Void of Empty Space" is to be understood in the sense that it references only the vastness of Space. Obviously, there is never a void of material in the Void of Empty Space. Also, the words "interstellar medium" (ISM) is defined as being 90% Hydrogen, 9% Helium, and 1% Dust. This is known as the rarified atmosphere of Space. To be sure, "Empty Space" is never empty of material.

Further as to wording referencing Dark Matter, just recently on 10/1/2017 while reading my very old paperback book LETTERS FROM THE EARTH by Mark Twain (copyright 1938) in the section entitled "Papers of the Adam Family" (as in Adam and Eve in the Garden of Eden), it is written that Adam and Eve are still in the process of learning about everything surrounding them for the first time ever! Their science conclusions come about by long investigations of mysteries that they see before them, and by studying the cause and nature and purpose of everything they come across. (Since there was no one to teach them anything, they had to figure things out for themselves.)

It was Adam that made the first scientific discovery, which was that "water only ran downhill, not up." Then later both Adam and Eve long since arrived at the conclusion that "atmospheric air consisted of water in invisible suspension." The words invisible suspension certainly caught my eye. It's as if my personal wording for Dark Matter/Pressure Ether could be stated as: **spatial air consisting of ether in invisible suspension**.

In my long investigations in searching for the CAUSE of Dark Matter among the five primary objects in the planetary Universe (Suns, Planets, Moons, Asteroids and Planetary Rings, Comets and Meteors), this simple process of elimination strongly suggests that Suns per se must be substantially contributing to the CAUSE of Dark Matter/Pressure Ether in some yet to be discovered way. Next came the process of deciphering the many activities of the Sun; one of which could be the cause of Dark Matter production. Spicules, Prominences, and Solar Flares of Suns were studied in depth. Which of these three ongoing activities could produce Dark Matter? Unfortunately, none of them is the answer. Why? The answer to that question is simply because there are vast regions of the Universe where Suns *do not exist* – but where Dark Matter does and must exist.

Even the luminosity factor of a star was considered. For example, our mid-sized Sun has a surface temperature of 10-11,000 degrees Fahrenheit (F.); Sirius B at 48,000 degrees F.; Betelgeuse at 4,900 degrees F.; and S Doradus is stated to be a million times more luminous than our Sun. However, luminosity as a whole from many different stars would only produce **variable**, out-of-sync results for Pressure Ether. This cannot be. This universal substance must be created by an ongoing sequence which stimulates a **steady**, reliable result in the Void of Empty Space.

Last it must be pointed out that there were two other concepts that I entertained as the CAUSE of Dark Matter which proved to be mistaken concepts; namely, 1) Rotation of Planetary Bodies, and 2) Amoeba Particles via Solar Wind.

1)   Regarding the rotation of planetary bodies to equate to giving off dust and energy that translates to Pressure Ether or "black matter" of the Universe, as it was referenced at that time long ago, here again the Void of Empty Space proved to be its undoing. There are vast regions of Space with the total absence of any planetary bodies. As a conclusion: "Even to visualize particles of matter inhibiting such Space, without the 'pressure' imposed by the rotation of these massive objects whereby the spark of creation for Dark Matter could take place, this sequence of events would be non-existent."

2) In June 2005, my notes asked: "Q. What if the ether (called also "black matter") that is in space between all planets and suns, etc., is a MASS that feeds off the nearby SUNS (its proximity to) and is in a CONSTANT STATE OF EXPANION?" In short, what of the possibility for a Sun to create a causative condition whereby Dark Matter is constantly being initiated by an amoeba-like action occurring in Empty Space requiring a constant flow of charged particles *from the Sun*, and the Solar Wind is responsible for dispensing said particles in its travels? Unfortunately, this Sun distributing particles via Solar Wind which would self-replicate in an amoeba-like action cannot be justified due to the cessation of the flow of Solar Wind; i.e., at some point Solar Wind will eventually terminate, thereby causing the necessary flow of

charged particles to cease. Bottom Line: There can be no blank pockets of the Universe which possess no Dark Matter.

ADDITIONAL SEARCHING FOR THE CAUSE

To further search for an entity-catalyst which exists on a grand scale in the Universe, one is left with the most prevalent sightings; namely, Nebulas, Suns, and LIGHT.

Nebulas are gaseous formations of many sizes and shapes which are basically compacted, although some appear to be blossomed in nature and therefore appear to be on the wane. Many Nebulas have been seen and named by astronomers; for example, the Eagle Nebula, the Crab Nebula, the Horse's Head Nebula; the Rosetta Nebula, the Veil Nebula, etc. The reason that they come in so many shapes and sizes would seem to be due to the Solar Wind in the region; i.e., this forceful mass wind most definitely would have the ability to shape gaseous clouds which we call Nebulas. Even so, these gaseous patterns are rather stationary for some length of Time.

Very briefly, the **accretion theory regarding the formation of Suns**, including left-over Planets and Moons, gives no explanation as to the genesis or eventual presence of any Nebula, which is needed for that theory to continue! These gaseous clouds are a very infrequently discussed subject in detail which is probably due to their unknown exact nature. Nevertheless, these structures are blindly presented as being the vast clouds of gas needed for the **birth of stars**; in which one vast cloud will collapse to begin a process taking many, many years to produce a nuclear fusion Sun. The ratio is one cloud equals one Sun. Consequently, this current theory still clings to the scientific community as it states: "Nebulas are the birthplace of stars." I adamantly disagree.

Nebulas are also reported 1) formed by a total destruction of a number of Suns; 2) that Suns are born out of Nebulas from previously dying Suns which exploded and produced the gas Nebulas initially; and 3) these clouds produce countless Suns whenever roaming particles intertwine or mix with the nebulous cloud (even though this flat statement is unexplainable in detail by the scientific community). And furthermore, to say that a nebulous cloud must mix with a **different set of select particles and the result is a Planet** is absolutely absurd. Can anyone say that Nebulas can select particles indiscriminately that would churn independently into a Planet, rather than a Sun? Both of these happenings defy logic. I totally disagree with any and all of these statements.

In any event, as for Nebulas being a causative factor for creating Dark Matter continuously throughout the Universe, these gaseous clouds simply become diluted and may dissolve over universal Time. Therefore, they can be dismissed for any relationship to Dark Matter.

Suns would be the next entity to investigate regarding producing Dark Matter/Pressure Ether in the Universe. Obviously Suns constitute these magnificent objects from the tiniest newborn star to the largest known Sun of VY Canis Majoris in the Universe. On first glance, it would appear that Suns would be the overwhelming choice for a relationship to Dark Matter. However, the nagging problem is that there can be vast, unimaginable distances between Sun locations. This simple fact negates any chance that Suns could be the sole producers of the all-pervading Ether.

LIGHT is the final remaining topic to explore in this series of searching for the CAUSE of Dark Matter. Historically, on November 9, 2006, I entertained the thought in my notes: Dark Matter ("Ether") What If: Light creates Ether. *end note* Now the question of how, when, where, and why needed to be dissected in earnest. Needless to say, my research notes on this subject are voluminous.

Finally, on December 28, 2006, my notes conclude: "As of today, I am going to call the force that makes planets push apart and which scientists are calling 'Dark Matter in the Universe' – my wording will be: PRESSURE ETHER." This wording is still my personal preference as it encompasses both aspects of the invisible, unknown substance which is the glue that holds everything in its place in the Universe and the resultant Ether Web that is the result of the build-up of Pressure Ether.

THE HYPOTHETICAL CAUSE OF THE ETHER WEB

Most important was one of my earliest thoughts: **"What happens when HOT meets COLD?"** In August 2008, a question in my notes: "What If: The HOT of the Sun particles HIT the frigid COLD of the temperature of empty space and PRODUCES A STEAM? This STEAM then SOLIDIFIES to a degree to form a BLANKET OF ETHER."

To envision that a web-like substance could be produced and maintained in Empty Space is relatively easy to picture. But the activity of HOT meets COLD should be explored. Perhaps the *weight* of the frigid background temperature of the Universe involved could solidify (for lack of a better word) the chafing action of the enormously hot LIGHT speed as it passes through the environment, and the resultant cold, cold Pressure Ether would then **gel** into place. An Ether Web could logically exist; not as a solid, but as a web or cocoon-type structure which totally permits huge planetary bodies to pass through it unhindered. Further still, because the constant temperature of Empty Space is constantly cold, there would be no chance for an Ether to ever be broken down or dissolved into a state of hot which would totally disrupt the entire chain of function in the Universe; namely, the web of Dark Matter holding everything in its place.

In May 2009 with my thoughts on Pressure Ether, it was envisioned that: 1) LIGHT (hot) is creating a friction, a constant chafing of the environment of Empty Space (cold) due to its tremendous speed of travel (although not exorbitant in terms of universal proportions); and 2) the Ether produced by the chafing friction and pressure of the speed of LIGHT will thin out or entwine, due to extreme HOT meeting extreme COLD. Therein lies the basis of my terminology: **the Ether Web**.

Now that the cause of Dark Matter has been directly linked to a Sun emitting LIGHT, several questions must be asked regarding the Sun per se, the emission of LIGHT, and the continuous creation of an Ether Web of Dark Matter. First, what happens when a Sun dies? If it shrinks and shrivels down to a dead state and ceases to be able to produce the hot substance of LIGHT needed to interact with the frigid surroundings of Empty Space, what happens to the production of the Ether Web? If this web of Ether has no food on which to sustain itself, what will happen to the immediate area of Ether volume? Needless to say, it's all well and good to theorize that the source of Dark Matter has been found by Suns emitting LIGHT, but when that source is no longer there, questions need to be asked and answered regarding the existence of the Ether Web.

1) What happens to the existing web of Ether surrounding a new dead Sun?

A. Whatever quantity of Pressure Ether/Dark Matter is present at the time of a dying Sun, the universal pressure density of the Ether will simply continue to mingle, because there are no other elements that exist within the emptiness of Space which can do it any harm. The chemistry of this product (as it can be presumed to be) is impervious to its surroundings.

2) What is the Ether Web's "shelf-life"?

A. Its shelf-life will never cease, but it will merely continuously intertwine with existing Pressure Ether.

3) Is the Ether self-perpetuating at some stage, as in an amoeba state?

A. Because it merely intertwines with existing Pressure Ether, consequently there is no need to self-replicate.

4) Will the Ether Web be reduced in size over time (universal Time) and thus causing certain areas of Suns and galaxies to move closer together?

A. Until the actual chemistry composition of the Ether can be known and using the assumption that the Dark Matter material cannot be tampered with, there will be no need to worry about any areas moving closer together due to the Ether Web being reduced in size.

5. Does the gravity-density of the dying Sun "draw in" an amount of the Ether Web at the Sun's central critical density point, thus reducing the quantity involved which in turn reduces the expansion of the Universe?

A. Because Dark Matter/Pressure Ether has no gravity, then the drama of a dying Sun will not affect any of the surroundings of an Ether Web.

6. Are the Ether contents of such a web structure not hampered by any gravitational pull of Suns, alive or dying, throughout the cosmos at any time?

A. Again, because Dark Matter/Pressure Ether has no gravity, the pull of any Sun with its gravity base will not affect its Ether Web surroundings.

THE HYPOTHETICAL CAUSE OF DARK MATTER

In December 2008, the simplicity and the majesty of it all was there to see: Speed of LIGHT traveling in the Void = Friction or chafing of the Void = non-stop creation of Dark Matter in the Universe.

This sequence of events was the absolute result of envisioning in a pictorial mind-experiment: **What happens when extreme HOT meets extreme COLD?**

In my March 2010 document, the following three principles were outlined for presenting what may be the sequence of events in producing Dark Matter aka Pressure Ether throughout the Universe.

1) Friction: the Catalyst

Friction is the key to the <u>cause</u> of the unknown substance existing as a buoyancy blanket (my terminology) that holds everything in the Universe in its place.

2) Speed of LIGHT: the Agent

The Speed of LIGHT is the necessary <u>action</u> which produces the friction within the Void of Empty Space.

3) Pressure Ether: the Product

Pressure Ether/Dark Matter is the never-ending <u>result</u> of what takes place when the Speed of LIGHT creates overwhelming friction or chafing while traveling at 186,172 miles per second through Empty Space.

Thus, constant LIGHT speed and the pressure involved are enabling a process to sustain itself. That is to say, a minuscule unknown chemical reaction is taking place in the Void. Dark Matter/Pressure Ether is a by-product of LIGHT which is constantly speeding non-stop through the environment of Empty Space. This never-ending action produces the friction catalyst in creating the buoyancy blanket Ether Web existing throughout the Universe.

A question was posed regarding this unknown action of LIGHT chafing the Void of Empty Space:

Q. If the wavelength of LIGHT generally determines its color, then why would Pressure Ether be invisible or having no other color except black?

A. The obvious reason is that the Ether is a stable, web-like oriented, compacted gas **unmoving at high speeds**; whereas LIGHT is radiated gas **moving at 186,172 mps** which creates a distance between these successive waves and produces an attainable color for physicists to categorize and evaluate.

In summary, because the primary function of Dark Matter/Pressure Ether is to exist as a buoyancy blanket in which all universal objects safely reside: **DARK MATTER HAS NO NEED TO TRAVEL AT ANY DISTINGUISHABLE SPEED.**

This overall universal substance at the present time is completely unknown as to its chemical make-up and subsequently unknown as to its exact allowance for all materials in the Universe to pass through it in a "non-hindering and non-binding" way (my terminology). Nevertheless, considering the infinite dimensions of the humanly impossible to conceptualize scale of the Universe, there remains the fact that it would not be possible for all planetary bodies to evolve smoothly IF there was not something holding everything in its place. Thus, Dark Matter/Pressure Ether is as vital to the Universe as are the life-giving Suns which continue to perpetuate the cosmos. As long as there is the creation of Dark Matter, there will be a place for all Suns and all universal objects to reside.

THE FIRST DARK MATTER FROM THE FIRST SUN

Note: Just to be perfectly clear, it is my belief that the First Sun of the Universe came into being from its Pure Hydrogen surroundings background and went forward in the creation of the Universe. On the other hand, physicists using the Big Bang Theory format for the creation of the Universe must repeatedly state: **"We don't know how the first Sun was born."**

The creation of the First Dark Matter from the First Sun of the Universe can be told as a *strictly hypothetical theory* or pictorial mind-experiment. It can be envisioned that the First Sun produced a web of Dark Matter/Pressure Ether substance which subsequently surrounded this magnificent structure long before any radiation/LIGHT actually **permanently left** the Sun structure. And as stated in my March 2010 document: "In the case of the initial, absolute creation of the elements, Gravity had no form, no function as of yet. Therefore, an initial necessary housing was needed for these elements to interact; a necessary buoyancy of Space. Thus, Pressure Ether/Dark Matter had to be a creation in the very beginning that the elements came into being." All elements, of course, were made **INSIDE the First Sun**. Thereafter, upon their first journey out from the Sun, they would have needed an ever-present housing, a casing, a buoyancy blanket for which to **be in; thus, Dark Matter**.

As a comparison, the Big Bang Theory states that a galaxy formed (sans Suns) after hundreds of millions of years' time span in which one universal plasma cloud finally thinned out after the big bang explosion, even though that plasma cloud was filled with only the <u>lighter elements</u>. Then it is flatly stated that the first Sun of the Universe is to have formed out of the galaxy cloud and embedded into the galaxy cloud. However, the actual process for a single Sun formation or Sun birth in the Big Bang Theory writings is never addressed (not even hypothetically), leaving physicists to make the above statement: We don't know how the first Sun was born. Then it is stated that this first Sun had to <u>explode</u> in order to scatter the <u>heavier elements</u> made inside it. BUT it is further stated that not all of the elements were made at this time due to the fact that the first Sun of this theory was declared to be *quote* **"inadequate"** *end quote*. It is then flatly assumed that second or third generations of <u>exploding Suns</u> would finally accomplish all of the element production tasks inside stars at some future time, somehow.

My vision for the creation of the cosmos has no exploding Suns.

Here again, the creation of the First Dark Matter from the First Sun of the Universe can be envisioned as a logical sequence of events. Once the creation of the First Sun had reached the stage of giving off its first radiation, it can be envisioned that this Light was unable to immediately leave and escape out into the Pure Hydrogen background from whence the First Sun developed. It would have been naturally thrust out from the First Sun in a natural activity, but it would have been pulled back down to the body of the Sun due to two simple reasons: 1) self-preservation and 2) lack of energy. Because of the initial lack of energy for an escape velocity, radiation would then be held into a cycle of continuously being jettisoned out into the surroundings of the Sun and immediately falling back down to the Sun structure.

With the consequences of this happening for quite some time in the life of the First Sun of the Universe, these actions would create the production of Dark Matter/Pressure Ether **for the first time in the cosmos**. It must be remembered that it is not the radiation (LIGHT) itself that is solely responsible for this production, but it is the Speed and Friction that causes Dark Matter to be created. The following sequence of events would have transpired:

1) Friction would be the catalyst naturally occurring as the radiation is being thrust back and forth from the Sun.

2) Speed of the radiation would be the agent of action which causes friction to take place.

3) Pressure Ether/Dark Matter would be the never-ending result of what occurs when friction and speed involved with LIGHT is being forced back and forth from the Sun on a continuous basis.

It can be envisioned that for quite some time the production of Dark Matter was an ongoing process. The First Sun would then have been surrounded by a wonderful Ether Web that has slowly intermingled with the Pure Hydrogen background in which the

Sun was residing, and these two substances would exist at the same time. It's quite clear that as the interstellar medium is defined as being 90% Hydrogen and that Dark Matter is said to inhibit every nook and cranny of the Universe, these two substances have been existing at the same time since the beginning of Time.

FIRST LIGHT'S JOURNEY

The absolute **First Light** of the Universe being emitted from the absolute **First Sun** of the Universe is a specific subject which has never been addressed in any creation theory to my knowledge. However, this subject is of vital importance in the development of the theory of the CAUSE of Dark Matter, and it has necessitated a pictorial mind-experiment to envision First Light's journey. Thus, the following is a hypothetical depiction of the answer to the question: Where would First Light travel "to"?

For the record, the thrust of Light off from our average-sized Sun at its 4.5 billion-year-old stage has an **escape velocity** of 617 kilometers per second, or 0.2% of the Speed of Light. Obviously when the First Sun of the Universe began to emit radiation, that pulse or thrust may have been tremendous, but as a beginning feature, it could be seen as having been **faint** (so to speak). This initial lack of energy for an escape velocity would then create a situation whereby it is ultimately maintaining its self-preservation. Consequently, all radiation for some time would be thrust out from the First Sun and fall immediately back down to its surface. It must also be remembered that at this beginning phase of the First Sun of the Universe is the fact that no newborn Sun cores, which would be the genesis of other Suns, have been created/born at this point. There is only the First Sun. It stands alone.

Using the philosophy of "LIGHT attracts LIGHT" in this instance, many questions come to mind. Could all photons ejected by a Sun (even though bound by some unseen law to travel in a straight line) be searching out another Sun? Is this their ultimate goal? Is it essential to find and be re-united with other planetary bodies producing Light? This never-ending production scenario leaves a trail of Light heading where? What are these corpuscles or photons seeking? Is this act of self-propelling energy all that is accomplished in a non-stop fashion?

The above questions may all sound trivial. However, First Light is a subject that holds many unanswered questions. It is this one main principle of attraction that holds the key to First Light's initial journey. That is, when the First Sun, over an unknown quantity of Time, had ejected many newborn Sun cores in all directions (my creation theory), these small objects of Sun material had to eventually develop into the time-consuming fusion stage and begin emitting radiation themselves. Only then could Light from the First Sun have an attraction to Light which it could travel "to" and thereby leave the grasp of the First Sun in a journey to a destination. (Note: All of this is possible even if there is a belief in a different version of the formation/creation/birth of Suns.) Again, any Light leaving the First Sun would be seeking other Suns due to the philosophy of LIGHT attracts LIGHT principle. How fast the production of new Suns came into being from the First Sun of the Universe in the Beginning of Time is, of course, an unknown. Yet, looking at the Universe of today, observers constantly see an astounding quantity of Suns in all directions of the cosmos which is stated to be on the order of from 10 to 25 billion years old.

In summary, it remains paramount that initial First Light emitted from the very First Sun of the Universe would have been pulled back down to the First Sun for a great quantity of Time **before** any birth of other Suns would have taken place. This magnificent structure stood alone as the beginning of the Universe. Its development would be the first step in the creation of the cosmos. Thereafter, First Light's journey as it pertains to the development of the First Sun can be seen in a very natural sequence of events.

ENVIRONMENTAL IMPACT

It is vitally important that the substance of Pressure Ether/Dark Matter be able to pass through all of the elements in the Universe, as well as all of the universal bodies present. That is to say that whatever this "gas, mist, steam, gel" consists of, it must not interfere with existing elements and/or temperatures that are established in the Universe as we know them today. Therefore, the following is given for 1) the Composition of the

Interstellar Medium; 2) the Temperature of Empty Space; 3) the Interplanetary Medium Temperature; and 4) the Chemical Composition of Air.

1) <u>Composition of the Interstellar Medium (ISM)</u>: The interstellar medium refers to the vast, almost incomprehensible space at times between stars. It has been determined to be 90% Hydrogen, 9% Helium, and 1% Dust. It is known as the rarified atmosphere of Space.

2) <u>Temperature of Empty Space</u>: The temperature in the Void of Empty Space is stated to be 2.725 degrees Kelvin, which is at -454.72 degrees F. or -270.4 degrees C. This temperature is very close to what is termed <u>absolute zero</u>, which is the lowest temperature when the movement of matter is said to cease. Absolute zero is stated to be -459.67 degrees F. or -273.15 degrees C. Please note that 2.725 Kelvin is presumed to be an <u>average temperature</u> in Space for the cosmic background radiation (CMB) deduction. However, the temperature for Outer Space would actually vary in different areas in Space, depending upon Sun activity in the region.

3) <u>Interplanetary Medium Temperature</u>: The interplanetary medium can be stated as the thinly scattered matter which exists between the Sun, Planets, and Moons of our Solar System. Because this medium in our Solar System of 8 planets includes dust particles, solar wind, and obviously the heat from the Sun, it is only stated that a wide temperature range can exist at various regions. For example, Planet Mercury is closest to the

Sun at 29 million miles at times, but Planet Neptune is the farthest from the Sun at an astounding average 2,795,000,000 miles (2.8 billion miles) distance.

4) <u>Chemical Composition of Air</u>: The atmosphere of Planet Earth, known as <u>air</u>, encompasses the planet and is held there by gravity. Earth's air is reported as containing 78.09% Nitrogen, 20.95% Oxygen, 0.93% Argon, 0.03% Carbon Dioxide, and very small amounts of 13 other gases. Air also contains an average amount of water vapor of 0.4% for the atmosphere itself. Scientists have labeled five main layers regarding air pressure and temperature; and there are five secondary layers called the ozone, ionosphere, homosphere, heterosphere, and the planetary boundary layer. As you can see, the air we breathe, as well as in relationship to the Earth, has been well defined. Needless to say, the 21% Oxygen factor is absolutely vital to human beings, and Pressure Ether/Dark Matter has not interfered with that situation.

The amazing thankfulness for this unknown gas/mist/steam/gel to have shown no signs whatsoever of intruding into or re-arranging any of the above elements or temperatures is truly mind-boggling. Scientists must keep this in mind in their quest for any knowledge as to its composition.

WHERE DOES DARK MATTER STOP?

The hypothetical theory herein has posited that the production of Dark Matter can rely on all Suns producing LIGHT as its causative point of beginning. However, the question begs: Where does Dark Matter/Pressure Ether stop, if anywhere, and why?

In a logical sequence of events, it can be envisioned that LIGHT from the First Sun of the Universe would at some point actually **leave** this fantastic structure and travel out forward into the surrounding Dark Matter which was created long, long ago. But just as in today's Universe, the catch-22 question would be: How can LIGHT travel beyond its contents in Space; i.e., can enough Dark Matter always be in existence so that LIGHT would never have to fear encountering an ending boundary to the Universe? More importantly, how could LIGHT propagate efficiently in its safe buoyancy blanket of Dark Matter if this all-encompassing substance ceased to exist? The answer can only be that the production of this substance in the Universe *must always stay ahead* of the radiation of all Suns in the Universe and, of course, ahead of all the creation of new Suns and Planets that will one day reside in this safe buoyancy blanket of Dark Matter. Theoretically, the thought that Dark Matter would always exist ahead of any future LIGHT speeding on its journey should not be a problem

The conclusion is that in the Beginning, the First Sun of the Universe produced a significant web of Dark Matter substance surrounding its magnificent structure long before any radiation actually left the Sun. The bounty, the scope, the dimensions to which a new Universe was being created with the vital substance of Dark Matter as an integral part: this, of course, is beyond human comprehension. Further, as for a "boundary of Space" existing in the cosmos – an actual ending region – this is truly beyond human understanding, and in my humble opinion, it would serve no purpose. The never-ending existence of Space and/or the nothingness Space of no known boundary should and must exist [SOMEHOW] in order to be filled with bountiful Hydrogen, which is the most abundant element of the Universe and which exhibits the attributes of forever expanding. In other words, **unbounded Space** must be constantly ready for habitation. The processes by which such an existence should be forever forthcoming are, of course, completely unknown.

As to the original question of "When does Dark Matter stop?" – my notes reflected in 2010 my envisioning that Dark Matter positively cannot function in an environment other than Empty Space, and therefore when it reaches the atmosphere of ANY planet or moon, it will cease to be. The buoyancy blanket will be there to hold the planet or moon in its natural place, but it will not "penetrate" any material body. However, with my purchase of OPTICKS by Sir Isaac Newton, I was compelled to write four pages of notes regarding Pressure Ether/Dark Matter that had never entered my mind.

After reading Query 21, it becomes quite clear that Newton is stating that AEther (21st century Dark Matter) **is in bodies of**

Suns, Planets, and Comets "at a much lesser rate" than in the Void of Empty Space. He makes no hesitation in placing this AEthereal Medium (Newton's terminology) as **blending into these dense planetary bodies upon encountering them**. Obviously, my thought that Dark Matter would cease to be in any planet or moon was totally incorrect. That is to say, one must bow to the genius reasoning and wisdom of Sir Isaac Newton.

Ironically, this is all envisioned even though Newton states quite candidly that he does not know what the AEther is! And what I find somewhat shocking is that from 1704 to the present day of 2017, no one else knows what Dark Matter/Pressure Ether or the AEther is!

For the sake of clarity, one can picture, as Newton envisioned, that the entering of this AEther into a dense body's atmosphere or surface or actual planetary material *could possibly* necessitate its being in a <u>much lesser rate</u>, diluted form, than would be absolutely necessary in Empty Space as the substance that holds everything in the Universe in its place. This leads to the final conclusion that as this buoyancy blanket encounters a dense body, it will surround, <u>as well as penetrate</u>, the entire entity. As for a "non-hindering" substance, all planetary bodies glide through this Pressure Ether via their rotations in Empty Space.

Note: The thought that a "gas/steam/mist/gel/ether" would penetrate a dense body so as to interact with its chemical make-up has never been addressed, to my knowledge. But as a major deduction, it would be logical to conclude that the AEther **does not interact** with any elements whatsoever; thus my terminology of "non-binding." Its chemical make-up must be *above and beyond* any known substance to Man. Nevertheless, it makes perfectly good sense to assume that **Dark Matter/Pressure Ether does not stop at planetary doorsteps**!

A final startling aspect of a Universe having Dark Matter being created by the speed of the emitted LIGHT of all Suns is the supposition that this tremendous non-stop production may be causing, quite naturally, an overload of Dark Matter to exist in the Universe as a whole. Consequently, this invisible, all-pervading universal gas product could be causing an expansion, however minute in scope, in certain regions between great galaxies, even though it has the ability to be non-hindering and non-binding to all that passes through it. There is no mystery as to what is happening. BUT there is absolutely no need to incorrectly theorize that a *quote* "75% Dark Energy" *end quote* exists because of and is caused by Dark Matter.

An Ether Web would have no need to possess great hordes of energy, and Dark Matter would have no energy per se except the sheer magnitude of the bulk it represents. As stated previously: Gravity is a force; not a substance. Electricity is a force; not a substance. Pressure Ether/Dark Matter is a substance; not a force. In any event, until the actual *composition* of the Ether is known, what effects this imposition has as it pervades onto all dense bodies in the Universe will remain an unknown; however, resting assured it is a benevolent encounter.

As can be seen, there is still so much to be learned regarding such a vital and integral part of the Universe as Pressure Ether/Dark Matter. This elusive subject will hold great attention for many future generations to come. New thinking is required.

Summary: The theory presented herein for the sequence of events in producing Dark Matter aka Pressure Ether throughout the Universe is given as follows:

1) Friction: the Catalyst. Friction is the key to the cause of the unknown substance existing as a buoyancy blanket that holds everything in the Universe in its place.

2) Speed of LIGHT: the Agent. The Speed of LIGHT is the necessary action which produces the friction within the Void of Empty Space.

3) Pressure Ether: the Product. Pressure Ether/Dark Matter is the never-ending result of what takes place when the Speed of LIGHT creates overwhelming friction or chafing while traveling at 186,172 miles per second through Empty Space.

Constant LIGHT speed and the pressure involved are enabling a process to sustain itself.   A minuscule unknown chemical reaction is taking place in the Void. Dark Matter is a by-product of LIGHT which is constantly speeding non-stop through the environment of Empty Space. This never-ending action thus produces the friction catalyst in creating the buoyancy blanket Ether Web existing throughout the Universe. * * *

CONTENTS: NEUTRINOS

NEUTRINOS

INTRODUCTION

The writing presented herein consists of defining and exploring a very popular 21st century subject; namely, the Neutrino. It was first described as the "ghost particle" due to its nature of being totally unable to be detected. As of today, it remains as such.

Let me be very clear that my interest in this topic had been zero during the years of 1999 to 2012 when my writing involved only the subjects of LIGHT and Dark Matter. However, in finding that the scientific community was completely focused on **capturing** Dark Matter, it was in later years that it became obvious that science was also dramatically focused on **capturing** Neutrinos.

Here again, this mysterious particle has resulted in many experiments towards determining just what this elusive particle is all about. My goal is to set forth a substantial amount of known facts at the present time and to then evaluate the results. As everyone knows, not all theories can be justified, and a good many have to be abandoned. Perhaps Neutrinos can be actually stated as fact, or on the other hand be dismissed as wishful thinking imposed in order to simply satisfy a basic law of science; namely, the Law of Conservation of Energy.

WHERE A NEUTRINO BEGAN

Before exploring the hypothetical characteristics attributed to the Neutrino, one must understand that the action event which produces this hypothetical particle begins in the Atom. Scientists now know that atoms are made up of three particles; namely, neutrons, protons, and electrons. The characteristics of the Atom are reported that neutrons and protons have structure and are held together by strong interactions, while electrons (and Neutrinos) can be envisioned as structureless and are extremely small objects that behave like a wave and/or a particle. Atoms are stated as being the basic units of matter. How the Atom per se came to be created in the Universe is still open to debate; as the creation of the Universe is also a topic of debate. Nevertheless, Atoms consist of three particles whose activity can produce "on paper" a new particle called the Neutrino.

It was only in 1911 (a scant 106 years ago) that Ernest Rutherford, a New Zealand physicist moved to England, proposed the nuclear model of the atom; i.e., the atomic nucleus. In 1920, he then proposed that a positively charged particle in the nucleus would be called the **proton**. (However, the proton was "discovered" by Wilheim Wien in 1898.) Rutherford also theorized that a neutral particle was within the nucleus, which was later discovered by James Chadwick in 1932 and would be called the **neutron**. Chadwick came to Cambridge and was second only to Rutherford as they worked together.

It was in 1897 that J.J. Thomson, a British physicist, discovered the **electron** in his work with cathode rays. He found negatively charged particles as he studied the "properties" of electric charge created in cathode ray tubes. As an elementary particle, which means there are no smaller parts in the composition of this particle, the electron has a negative electric charge (-1) that is <u>equal in magnitude</u> to the positive electric charge of the proton (+1) to which they are electrically attracted. These negatively charged electron particles are extremely lightweight and simply reside <u>in a cloud</u> in an orbital fashion to surround the atom nucleus. On the other hand, protons and neutrons are much heavier than electrons, and they reside <u>in the center</u> of the atom nucleus.

Electrons are identical as far as scientists know, and they are extremely low in mass. When Rutherford proposed that every atom was composed of a tiny <u>nucleus</u> which was said to be in the center of the atom, this accounted for 99.95 percent of the total mass, and the rest of the atom was made up of electrons in the outer regions. One calculation states that the mass energy of an electron as being merely 0.000548. Further, electrons can be "thrust out/kicked out" of any atom by certain processes, and one or more of them can be transferred from one atom to another. There is no set amount of electrons that an atom must maintain.

Within the atom, there is a process that occurs called a "Beta-decay event" in which a Neutron **converts** to a Proton and an Electron and a hypothetical Neutrino. When this event suddenly takes place, a so-called "beta particle" is thrust out of the atom and this beta particle **is the electron**. The wording "beta particle" is simply to distinguish that this particular electron was created in a Beta-decay event. Again, **a beta particle is an electron**.

Thus, there are three players involved in which an event occurs within an atom and a "left-over" particle (my terminology) is said to come into being, which is the topic of this writing; namely, the **neutrino**. The three players are: an original Neutron, a newly created Proton, and a newly created Electron.

Note: Scientists have gone into a strange field called **quarks** since 1964. Even though quarks, as determined to be elementary particles, have not been directly observed, it has been announced that neutrons and protons *quote* "are made of quarks" *end quote*. Specifically, neutrons are made up of 1 "up" quark and 2 "down" quarks; each proton is made up of 2 "up" quarks and 1 "down" quark; and they are all held together by other particles called gluons, which have also never been observed.

After reading and researching on this subject of quarks many years ago, my thoughts have not changed. That is to say, this subject remains beyond my ability to fully comprehend, and therefore it will not be fully addressed here except as to their reference with an event (a Beta-decay event) in the production of a neutrino.

DEFINITION OF 3 MAIN PARTICLES

In order to understand just how the Neutrino came into being, it is necessary to know the definition of the various actors involved; i.e., the particles which play a decisive part in producing an event which leaves a left-over particle or commonly called "ghost particle" behind.

Neutron: an unstable neutral subatomic particle which has no electric charge (0). It is said to have a "rest mass" which is nearly 1,839 times greater than that of an Electron.

Proton: a stable subatomic particle that has a positive electric charge (+1). Its positive charge (+1) is equal in magnitude to a unit of an Electron negative charge (-1). It is said to have a "rest mass" which is 1,836 times the mass of an Electron.

Electron: a stable subatomic particle that carries a negative electric charge (-1). It is said to be nearly massless in comparison to a Proton or a Neutron.

These 3 main particles are the items that form the atom. The **Neutrons** and **Protons** are the much larger particles located in the center, and the **Electrons** are defined as fast little energetic bundles of energy residing in a cloud which surrounds the entire nucleus containing the two larger particles. The mass of the atom itself has been described as having an internal structure of Neutrons and Protons that make up 99.95 percent of the atom, and the Electrons make up the rest in their outer cloud around the nucleus in orbital positions.

The speed of an Electron needs to be explained as clearly as possible. It must be remembered that Electrons act in a <u>wave-particle</u> duality role. They do not have a definite position within an atom; i.e., electrons can exist at <u>all</u> locations around the atom nucleus. That is why they are said to be found in "orbitals" but not "orbits." And the Electron speed at every point of any location can obviously be completely different in its relationship to the nucleus. Therefore, to determine a speed of Electrons as a whole is not a solid subject ready for definition.

Nevertheless, a calculation was found which stated that the Electron "at the fastest in the Hydrogen atom" is traveling at 2,200 km per second or 0.7% the Speed of Light. In other words, the Electron in this situation travels at an incredibly slow velocity in comparison to the Speed of Light (SOL). (There is no way to verify this mathematical figure and is only presented as a printed figure here.)

But again, Electrons do not have a definite position because they reside in a so-called <u>cloud</u> or <u>haze</u> around the nucleus, and consequently there can be no particular speed attached to them individually or as a group.

Finally, the definition of the hypothetical Neutrino can be stated as follows: <u>Neutrino</u> = an incredibly tiny subatomic particle produced in a Beta-decay event to explain where some "missing energy" has gone to; it is said to have hardly any mass to no mass whatsoever; it is not definitively detectable because it is believed to have no interaction whatsoever with any known item; and it carries no electric charge (0). For all of these characteristics, it was declared to be the "ghost particle" upon its creation.

NEUTRINOS FROM WHERE?

Neutrinos seem to have taken on a life of their own in my humble opinion. That is to say, there is a plethora of information about this left-over particle as originally theoretically presented during Beta-decay in the atomic nucleus. One would think that this is the only way that neutrinos can come into being. Needless to say, this is the original way in the early 1930s in which a Neutrino was depicted as the vehicle for the "missing energy" determined to occur in a Beta-decay event.

However, before detailing what happens in the atom when a Beta-decay event takes place, there are now several other ways in which neutrinos are now sought to be found. It's important to clear the air regarding these additional production procedures that **claim** to be a method to find the Neutrino particle. The following will list briefly some of the ways that have transpired with regards to finding this elusive hypothetical particle.

1. In the 1950s, in a nuclear reactor using Beta-decay by Clyde Cowan and Frederick Reines, an experiment was conducted whereby neutron capture and a positron (the electron anti-particle) destruction gave a result: an anti-neutrino interaction. This anti-neutrino emitted by a nuclear reactor is said to be the

"anti-particle of the electron neutrino." Comment: No neutrinos were ever detected.

2. In 1962, L. Lederman, M. Schwartz, and J. Steinberger put forth the supposition that a <u>muon-neutrino</u> was an interaction. The muon is considered a charged *lepton* which has a mass 200 times greater than the electron. This muon-neutrino is stated as a type or "flavor" of neutrinos. Comment: Any experiment to produce these leptons has never resulted in the detection of an actual neutrino. The neutrino is *assumed* to accompany it.

3. In 1975, a <u>tau-neutrino</u> was put forth at the Stanford Linear Accelerator Center. The tau is considered a charged *lepton* which has a mass 3,500 times greater than the electron. This tau-neutrino is stated to be a type or "flavor" of neutrinos. Comment: Any experiment to produce these leptons has never resulted in the detection of an actual neutrino. The neutrino is *assumed* to accompany it.

4. <u>Oscillation</u> by neutrinos was put forth and implied as a way to determine "neutrino flavor changes in flight" involved with solar neutrinos. Further, since neutrinos have all the attributes of a Photon, meaning it is uncharged, massless, and travels at the Speed of Light, scientists have even given it a Spin Number in the assumption that they are both spinning particles. The Photon has a spin of +1 or -1, and the Neutrino has a spin of +1/2 or -1/2. Comment: It must be remembered that neutrinos are said to travel at the Speed of Light. This wishful thinking by neutrino oscillation that a type of solar neutrino can be implied

is simply nonsense, in my humble opinion. Beta-decay neutrinos are said to be nearly massless and have no electric charge. What kind of oscillation can be expected in such circumstances? None. And further if these nearly impossible to detect neutrinos travel at the Speed of Light (with no interaction whatsoever), then no oscillation can be remotely possible. To be sure, no neutrinos have ever been detected using this method of deduction.

5. Observation of neutrinos from a particular supernova was put forth as a detection method. This would be due to neutrinos being able to pass through all of the material in an exploding star and fly out into Space. Comment: To use a supernova (which is actually a very rare occurrence in the Universe; estimated 3 will occur per 1,000 years) as a means to explain neutrinos escaping into Space is highly speculative. Nevertheless, even if you do believe in supernovas occurring in the Universe, no neutrinos will ever be detected from the remnants of a very infrequent exploding Sun. The technology is just not there.

As a summary, it is quite obvious that neutrinos per se are actively being pursued on an INTERACTIVE basis with various other items since they cannot be detected directly at the present time. This leads, of course, to a potential of over-estimating and over-inferring the actual deduction of a neutrino presence in an experiment.

NEUTRINOS FROM THE SUN

In the 1960s, the Homestake mine experiment in South Dakota proposed a hypothetical that the number of neutrinos from the core of the Sun was different as predicted by the "Standard Solar Model." These neutrinos are said to be created from the nuclear fusion decay rates within the Sun. However, a decay rate came to be shown as to vary at times; i.e., the decay wasn't constant. The varying decay rates is now focused on the so-called **solar neutrinos**. These are the massless and no electric charge particles that are theorized to be created as the fusion process of the Sun goes forward. These tiny trillions of left-over particle solar neutrinos in the process are now being cited as the cause of any varying decay rate transpiring in the Sun. This is still all speculation and unresolved.

Strange as it may seem, the decay rate was proposed as being connected to the SEASONS; i.e., producing a rate slightly slower in the summer and slightly faster in the winter. This phenomenon is still being explored, as well as a decay rate associated with Solar Flares.

[A list of neutrino experiments can be found on Wikipedia computer site (as viewed on 10/27/2017) which states that this is a non-exhaustive list of neutrino experiments, neutrino detectors, and neutrino telescopes; 51 names are on the list.]

It has even been calculated that "about 2 percent" of the Sun's energy is carried away by neutrinos created in the fusion reaction activity in the depths of the Sun. Common sense nevertheless would dictate that *anything* in the Sun has a profound effect on the Sun's energy output.

However, as for the "about 2 percent" of the Sun's energy being carried away by neutrinos (and even stated in other articles as being "perhaps 5 percent"), this makes no sense, in my humble opinion. Since when do neutrinos with no electric charge have such profound energy output? One can envision that tiny, tiny, tiny neutrinos on any voluminous scale would not have the capability to ALTER the function of energy output from the Sun. Their size and their characteristics would seem to negate any such effect.

Certainly there is no dispute that the <u>energy of the Sun</u> is brought forth out into the Universe in the form of **photons**. And it is reported that their journey from the center of the Sun is through over 400,000 miles of solar matter to the surface. And it is reported that photons are <u>easily absorbed</u> by surrounding material, once they are produced in the center of a 15 million degree Centigrade Sun. And even though photons are stated as being massless and neutrinos are stated as being massless, there can be no comparison of these two items. They are worlds apart!

Finally, the topic of **solar neutrinos** is a very popular subject in the 21st century. It is currently reported that **trillions of neutrinos pass through your body every second** and keep on going with no harm to you or to the Earth which they pass through onto a never-ending journey in Space! This whole thought is, of course, mind-boggling. Also currently reported is that a "solar neutrino flux" for us on Earth is **about 65 billion neutrinos passing through just one square centimeter of area on Earth every second!** Those kinds of figures are incredibly hard to comprehend. My only question is: What purpose would solar neutrinos serve in the Universe?

A conclusion can be logically reached that perhaps the concept of solar neutrinos may just fade from the scientific community in the future. At any rate, it must be conceded that with today's prevailing belief in this subject, these trillions of invading neutrinos have remained the most unharming and non-intrusive particles of all time.

BETA-DECAY TERMINOLOGY

The concept of neutrinos originally took hold in the early 1930s with the proposed dilemma that "missing energy" occurred in a process known as a Beta-decay event which takes place in the atom involving the breakdown of a Neutron. This Beta-decay, as a radioactive decay event, occurs when a Neutron converts into a Proton and an Electron. However, the concept of a hypothetical tiny particle was needed to be created in order to account for a minute amount of "missing energy" which developed during the event process. This newly envisioned particle came to be known as the NEUTRINO.

The following is the terminology associated with a Beta-decay event in which a Neutron converts into a Proton and an Electron.

Beta-decay = a type of radioactive decay involving the emission of Beta particles (Electrons).

Beta particle = an actual Electron originating in radioactive decay. This term can be confusing, but it is used extensively to indicate that this particular particle was created due to a Beta-decay event. A Beta particle simply can be envisioned as an Electron that is being thrust out of an atom immediately after it was produced.

Radioactive decay = the decay event in which a radioactive element spontaneously emits an energetic particle; usually an alpha particle, or a beta particle, or a gamma ray, or some combination of these items. Therefore, Beta-decay is simply one type of radioactive decay.

The stability of the 3 main particles of an atom is shown as follows:

1. Electron = little mass – it is stable – does not break down – negatively charged.

2. Proton = relatively massive – it is stable – positively charged.

3. Neutron = relatively massive – it is unstable – has no electric charge.

The characteristics of the new terminology of a "neutrino" are as follows:

Neutrino = no mass (or hardly any mass) – it has no interaction whatsoever with any known material – it carries no electric charge. These characteristics were developed and decided upon much later than when the neutrino was first conceptualized.

Further important terminology involved in a Beta-decay event is as follows:

Quarks = elementary particles which **make up** Neutrons and Protons. Quarks, as a concept, came into being in 1964. They are categorized into 6 types; however, in the Neutron and Proton, the only two that apply are the Up Quark and the Down Quark. In a Neutron which is made up of 1 "up" (U) and 2 "down" (D) quarks, and in a Proton which is made up of 2 "up" (U) quarks and 1 "down" (D) quark, there is a Beta-decay event happening by a weak nuclear force involving a **W particle (boson)**.

Virtual W particle (boson) = a subatomic particle with positive (W+) or negative (W-) charge that brings about the weak nuclear force. **These bosons change particles they interact with into other kinds of particles**. Notice that the **W** in this term represents the **w** in the word weak.

Note: To be perfectly clear, what is happening inside the nucleus can only be determined in an indirect way. That is to say, one cannot actually look inside the nucleus. The statement that a Neutron is changing into a Proton as an Electron is being thrust out of the atom cannot be <u>seen</u>. These determinations have been concluded over the years by scientists dedicated to understanding the internal structure of the ATOM since the 1890s.

RADIOACTIVE ENERGY IN THE ATOM

In the very late 1890s, Ernest Rutherford (1871-1937) became interested in searching for radiation which would <u>accompany</u> radioactivity. He was aware that in 1898, Marie Curie (1867-1934) was working to isolate radium. Shortly thereafter, Marie and her husband Pierre Curie (1859-1906) were awarded the Humphry Davy Medal for this discovery in chemistry, and in 1903, they shared a Nobel Prize with Henri Becquerel (1852-1908) who discovered the spontaneous radioactivity of uranium.

Rutherford's first discovery was regarding uranium radiation. In an experiment using layers of aluminum foil to block the radiation's travels, two things happened: one type of radiation was easily blocked and one type was penetrated easily. He named the non-penetrating rays **"a rays"** (later termed Alpha rays/particles). He named the penetrating rays **"B rays"** (later termed Beta rays/particles). At a later date, he discovered a third radiation activity caused by radioactive elements which he named **"Y rays"** (subsequently to be known as *not particles*, but more like x-rays and termed as gamma rays).

The Alpha particle was determined to be identical to the nucleus of a helium atom, consisting of 2 protons and 2 neutrons. The Beta particle was determined to be the same as an electron or its positive version called a positron.

In 1902, Ernest Rutherford and Frederick Soddy (1877-1956) published papers regarding their theory of transmutation of radioactivity (re elements). Then they studied and outlined what happens to radium, thorium, uranium, and actinium via radioactive disintegration.

Rutherford and Soddy also reported: 1) that radioactive energy was emitted from <u>within an atom</u>; and 2) that when Alpha and Beta particles were emitted, this event caused a <u>chemical change among the elements</u>. And at the time of a disintegration process, most involved the emission of one or more of the above type of rays; i.e., Alpha, Beta, or gamma. This theory, of course, was highly new and highly skeptical, but eventually it proved to be very sensible and was accepted as such. Thus, energy breakdown distribution was a very energetic field of study. Consequently, the Law of Conservation of Energy in the 1920s stood fast and foremost in science. This applied to subatomic particles as well.

1911: Rutherford described his "nucleus model of the atom" in which there is an electric charged center. He theorized the diameter of the charged nucleus to be: 10^{-13} centimeter. The size of the complete atom was theorized to be: 10^{-8} centimeter.

His model was initially not greeted with enthusiasm and was generally overlooked in the scientific community.

CONCEPT OF A NEUTRINO

In Beta-decay, the process was originally viewed whereby it was constant and would end up with energy going to the Beta particle (Electron) and with energy going to a new nucleus (a Proton). Therefore, everything went smoothly as an even-steven proposition or as a totally separated energy packet into two sources. Everything was accounted for.

In 1919, Lise Meitner had arduously climbed up into the physics academic world and became the first woman in Germany to have the title of Professor at what was then the radioactivity section of the Hahn-Meitner Laboratorium in Berlin. Then in 1925, Charles Ellis at the Cavendish Laboratory wrote about the work of Meitner regarding a mystery about Beta-decay; namely, what if the two sources of energy did not add up to a necessary sum total at the end of this event, and what happened to the remainder of the energy? The two scientists continued to stay in touch in order to solve this dilemma, but to no avail.

In the late 1920s and early 1930s, physicists were essentially dumbfounded and terribly annoyed by Beta-decay particles in regards to their energy activities. Simply stated, it was thought that when a Neutron breaks down to a Proton and then an Electron, some energy could not be accounted for and located.

Unfortunately, this means that the Law of Conservation of Energy would be in jeopardy which requires an even-steven energy situation. Q. Where is this missing energy?

In 1930, Wolfgang Pauli wrote a letter to the attention of Lise Meitner and Hans Geiger in which he simply stated his thoughts (his mind-experiment) regarding Beta-decay breakdown. Very briefly, he proposed that if the Electron in a Beta-decay event did not carry off ALL the energy, it MIGHT BE because a NEW PARTICLE would have been involved that carried off the REMAINDER. Therefore, a new additional particle must be involved in the breakdown event.

Pauli called this new particle a **neutron**, which he envisioned as being a neutral electrically charged particle. [Note: In 1932, the word **Neutron** came into being as meaning the actual large **Neutron** particle of an atom, and Pauli's word of a very small additional particle created would have to be changed to **neutrino**.] At that time, he further explained that in Beta-decay, a neutron (Pauli's word) would be ejected from the nucleus at the same time that the Electron was ejected. He proposed that they both must be on the same mass magnitude; which is very small indeed. Therefore, by having the concept of a very small new neutron particle (Pauli's words) existing in the nucleus, Wolfgang Pauli was able to solve the missing energy problem as to the Law of Conservation of Energy principle.

In 1934, Enrico Fermi proposed his theory of Beta-decay which included James Chadwick's actual named **Neutron** particle observed in 1932. These large **Neutron** particles were actual particles <u>in the nucleus,</u> along with the large **Protons**. Fermi envisioned that in a Beta-decay event, a Neutron was converted into a Proton, an Electron, and a Pauli <u>neutron</u>. It was then that Fermi decided to change Pauli's wording into a new word **neutrino** which means "little neutron" in Italian. This is how the word came into being.

However, Fermi believed that this **neutrino** is <u>created in the Beta-decay process rather than existing within the nucleus.</u> This theory proposed by Fermi ended the so-called loose ends of Beta-decay and solved a major problem for the Law of Conservation of Energy in this event. This is the theory that is held today; i.e., **an Electron is created and a Neutrino is created which both did not exist within the nucleus.**

It can be said that both Pauli and Fermi proposed that the "missing energy" in a Beta-decay event could be accounted for by simply adding a newly created particle.

As of this writing in 2017, neutrinos per se have never been captured and therefore have never been dissected or evaluated since their introduction into the scientific world in the early 1930s by two very esteemed physicists. This was a particle that existed "on paper" as an outlet when a certain event occurred. The ongoing speculation regarding these neutrino particles, which some scientists now want to state as a factual possibility, is still a theory to be yet proven to have merit.

BETA-DECAY WITHOUT QUARKS

The process of a Beta-decay event in an atom was known in the early 1900s; i.e., a Neutron breaks down and is converted into a Proton and an Electron. Then the Neutron disappears and the Proton takes its place. Then the newly created Electron is immediately hurled out of the atom to become in name a Beta particle. It's as simple as that.

$$\mathbf{n} \text{ converts to } \mathbf{p}^+ + \mathbf{e}^-$$

Further, any Beta particle produced should naturally now have only one amount of <u>kinetic energy</u> from this process. Kinetic energy is the energy of an object due to its motion. *But* if the nucleus involved which houses the breakdown Neutron is not even and always in the same state of being as to energy, then any kinetic energy involved in the process will vary; i.e., there will naturally be different kinetic energies that must be dealt with during the Beta-decay event.

Scientists then declared a "maximum value" for any kinetic energy. (Note: This author is assuming that this kinetic energy maximum value was calculated on size and spin of the Neutron.)

The MASS of the three players in a Beta-decay event comes into the Neutron breakdown process. They are as follows: Neutron Mass = 1.6749×10^{-27} kg. having no electric charge (0). Proton Mass = 1.6726×10^{-27} kg. with positive charge (+1). Electron Mass = 9.1×10^{-31} kg. with negative charge (-1).

Finally, it was known that emitted Beta particles lose an amount of MASS in the decay event, *but* this lesser MASS should be taken up with the kinetic energy maximum value rate. And it was known that no Beta particle ever exhibited a kinetic energy that was <u>greater than the value of the lesser decreased MASS</u>. Therefore, the Beta particle produced in a decay event was seen to **incorporate the kinetic energy** produced in a decay event. Thus, the Law of Conservation of Energy was accomplished and satisfied.

But there arose one very perplexing dilemma. Beta particles that were newly created in a decay event and hurled out of the atom possessed less kinetic energy than they should. That is to say, these new particles <u>barely even reached the kinetic energy maximum value</u>.

It has been stated that the energy was always less. It was so less as to be evaluated to be: 1/3 of the maximum. If so, this leaves a huge amount of the kinetic energy maximum value to be unaccounted for and missing. Consequently, the Electron (the newly created Beta particle) that is hurled out of the atom is **not accompanied with the so-called "missing energy."** My naïve question: Could it be the speed at which the Electron is thrown out of the atom that fails to incorporate any and all of the energy involved?

This is when in the early 1930s, Wolfgang Pauli and Enrico Fermi both presented different theories which held that a new particle could be created to incorporate this missing energy, and the particle was called the Neutrino **(x)**.

n converts to $\mathbf{p}^+ + \mathbf{e}^- + \mathbf{x}$

Thus, the Law of Conservation of Energy would be accomplished and satisfied.

BETA-DECAY WITH QUARKS

As stated previously: Scientists have gone into a strange field called **quarks** since 1964. Even though quarks, as determined to be elementary particles have not been directly observed, it has been announced that Neutrons and Protons *quote* "are made of quarks" *end quote*. Specifically, Neutrons are made up of 1 "up" quark and 2 "down" quarks; each Proton is made up of 2 "up" quarks and 1 "down" quark; and they are all held together by other particles called gluons, which have also never been observed.

Consequently, quarks are very much involved in the breakdown of one huge Neutron in a "Beta-decay event." Quarks are categorized into six types. However, in the Neutron Beta-decay event, the only two that apply are the Up quark and the Down quark. In a Neutron which is made up of 1 Up (U) quark and 2 Down (D) quarks, and in a Proton which is made up of 2 Up (U) quarks and 1 Down (D) quark, the Beta-decay event now happens by a "weak nuclear force" or interaction of a **Virtual W particle (boson)**.

This W particle is a subatomic particle written **W+** or **W-** to indicate a positive or negative charge. These particles are called "force-carrier particles" because they change particles that they interact with into other kinds of particles.

Needless to say, a Neutron Beta-decay event involves a lot of activity going on at one instantaneous time. The principle players are: 1) the original Neutron which will break down into a Proton; 2) a Down quark (D) and an Up quark (U) which are involved in the action; 3) the Virtual W-particle which brings about instant decay; 4) a newly created Proton which takes the place of the Neutron in this action; 5) a newly created Electron emerges from the W-particle; and 6) a newly created anti-neutrino particle (hereinafter simply referred to as Neutrino) which emerges from the W-particle.

Simply stated: A Neutron Beta-decay event occurs when a Neutron (UDD) decays to a Proton (UUD), an Electron, and a Neutrino.

Let's put this activity into a sequence of events as follows:

Note: The Neutron of UDD equals 1 up quark and 2 down quarks or 2/3 and -1/3 and -1/3. Total zero (0) charge.

Note: The Proton of UUD equals 2 up quarks and 1 down quark or 2/3 and 2/3 and -1/3. Total 3/3 or +1 charge.

1. One of the Down quarks (D) in the Neutron (UDD) **decays** into an Up quark (U) which makes a Proton (UUD). A Virtual W-particle is **emitted by the D quark of the Neutron**. The short-lived W-particle is called a "force-carrier particle" because it is responsible for bringing about the particle decay.

2. The newly created Up (U) quark springs back from the emitted force-carrier W-particle and takes the place of the Down (D quark) in the Neutron of (UDD) to become a Proton of 2 Up quarks and 1 Down quark (UUD).

3. The Neutron, as a whole, disappears as the Proton is made to take its place.

4. The decay or breakdown process has been "mediated" or brought about by the Virtual W-particle which then carries away a -1 charge. Thus, the charges involved are conserved or evened out.

All of the above transpires in a fraction of a fraction of a second.

5. The work of the W-particle is not complete yet. An Electron (and a Neutrino if believed) **emerges from the W-particle**. Then the W-particle decays/deteriorates instantly as its energy is transformed into the Electron and the hypothetical Neutrino. The work of the Virtual W-particle has happened in an incredibly short-lived time.

6. The Electron is an elementary particle which carries a negative charge **equal** in magnitude to that of the Proton's

positive charge. They both measure: 1.6022×10^{-19} Coulomb. The first "elementary particle" or the Electron was discovered by J.J. Thompson in 1897. The discovery of the Proton was by Ernest Rutherford in 1917.

7. The Electron from the W-particle (the force-carrier particle) then flies out and away rapidly from the atom, and it is now called a "Beta particle."

8. A (hypothetical) Neutrino particle is considered a left-over item and moves away from everything supposedly at the Speed of Light.

This entire Beta-decay process is stated as:

$$\mathbf{n} \text{ converts to } \mathbf{p}^+ + \mathbf{e}^- + \mathbf{v}$$

Actually, the above figure (**v**) is shown mostly with a line above a capital V and sometimes with a sub figure of e and refers to the "anti-neutrino" figure. In any case, the last figure in this formula above represents the left-over particle Neutrino.

Note: This writing does not specifically state the words "anti-neutrino" versus "neutrino" simply for the sake of simplicity. Technically, the Beta/minus decay of this writing deals with a neutron into a proton, an electron, and an "anti-neutrino." There is also a Beta/plus decay which deals with a proton into a neutron, an electron, and a "neutrino." But in this writing, all left-over particles are referenced simply as a **Neutrino**.

Note: There is the famous drawing by Richard Feynman in his book <u>QED</u> on page 140 which shows the interaction of the Virtual W-particle as it completes its tasks in the Beta-decay event. This drawing can also be seen on any good computer site featuring the workings of the Virtual W-particle. [QED, The Strange Theory of Light and Matter, by Richard P. Feynman, 1985, Princeton University Press]

In summary, the interactivity of all of the above will happen in such incredible fractions of fractions of a second and, of course, will not be able to be observed. Physicists are left with the understandings of what just transpired. They are to be commended for their diligence and patience in their endeavor to completely solve the Neutron Beta-decay event. That is why it is necessary to talk about the "missing energy" which still prevails in the above activity and perhaps to put forth some choices for a complete solution to this dilemma.

MISSING ENERGY MASS SOURCE

The Law of Conservation of Energy in the 1920s stood fast and foremost in science. Therefore, in a Neutron Beta-decay event where the Neutron converts to a complete Proton and into an Electron which becomes a Beta particle in name when it leaves the atom, all the energies should be accounted for. But this was found to not be so. That is to say, when energies involved with the new Proton and the Beta particle – aka the speeding Electron hurled out and away – were measured, there was a so-called "missing energy" value that needed to be accounted for and placed in a proper position. How could there be a missing energy between all of the Beta-decay event players?

Before any solution can be applied, the most important question has to be: Where did any missing energy originate from; i.e., what is its source? The only active players in the Beta-decay event are the original Neutron **(n)**; the two quark components of a Down quark and an Up quark; the weak carrier force of the Virtual W-particle; the newly created Proton **(p^+)**; the Electron **(e^-)**; and a hypothetical Neutrino **(v_e)**.

The Neutron Beta-decay process is stated as:

$$\mathbf{n} \text{ converts to } \mathbf{p^+} + \mathbf{e^-} + \mathbf{v_e}$$

Remember that it was Einstein who said that MASS equals ENERGY.

This Neutron/quark breakdown event is brought about by a Virtual W-particle, which is a force carrier that is **emitted from the D quark of the Neutron**. This W-particle is responsible for: 1) the Neutron Down quark changing to an Up quark; 2) carrying away a -1 charge regarding the quarks; 3) having an Electron and a Neutrino (technically an anti-neutrino) **emerge from** the W-particle itself; and 4) decaying instantly after all of these tasks have taken place. The very important Virtual W-particle is accounted for as far as its tasks.

Needless to say, this Virtual W-particle has performed many functions in this Beta-decay event. Consequently, there is a distinct possibility that some missing energy has come about somehow during and/or after all has been accomplished by this W-particle carrier force. The energy MASS sources involved in a Beta-decay event are stated in the following paragraphs, as well as the "plus or minus charges" of these players.

The Neutron containing a Down quark has a basic MASS energy value of 1.008665 a.m.u. (or 1.00867 in 5 figures). The small letters stand for atomic mass unit. The Down quark has a -1/3 charge. This Neutron, after it converts/disintegrates into a Proton and an Electron, will simply disappear. All of it is gone. All of it has been accounted for.

In the conversion process, one of the Down quarks in the Neutron (UDD) decays/transforms/converts into an Up quark that takes its place as a Proton (UUD). The Down quark is gone. The Up quark is part of the new Proton. All is accounted for.

The newly created Proton has a basic MASS energy value of 1.007276 a.m.u. (or 1.00728 in 5 figures). This Proton has a +2/3 charge. A W-particle carrier force, which has been **emitted from the D quark of the Neutron**, is responsible for the conversion of the Down quark of the Neutron into an Up quark for the Proton. The Neutron has decayed. The Proton has taken its place. All of it is accounted for.

The newly created Electron has a very small MASS energy value of 0.000548 (or 0.00055 in 5 figures). This Electron has a -1 charge, and it is said to **emerge from the W-particle** instantly. The Electron will then be instantly thrown out and away from everything. Because it is hurled out, it is now known as a "Beta particle" in name. This particle will go on its way. It is not a part of either the Neutron or the Proton.

However, because this Electron came from the W-particle carrier force action, there is a distinct possibility that some missing energy may have come about somehow after all has been accomplished and the W-particle has disintegrated. All *may not be accounted for.*

Finally, after all has been measured and added or subtracted in a mathematical calculation of the whole Beta-decay event, there is a missing energy dilemma which has been given an actual number. One will read that this missing amount has been determined to be: 0.000291.

That figure is obtained by taking the Neutron MASS figure and subtracting the Proton and Electron MASS figures. The Neutron MASS figure is: 1.008665. The Proton and Electron figures: 1.008374. The missing amount is: 0.000291. To be noted is that if this is the amount of the new so-called Neutrino, it would be *about* 1/2 an Electron MASS at 0.000274 (with a complete Electron MASS figure of 0.000548).

This tiny missing energy of 0.000291 among all of the players in the Neutron Beta-decay event **must be placed somewhere**.

Therefore, a new particle was invented/theorized "on paper" to be the recipient for this energy; i.e., the Neutrino of the early 1930s of Pauli and Fermi. Thus, the Law of Conservation of Energy was accomplished and satisfied.

Even though this Beta-decay event was seen as "something somewhere is losing energy" – which is the conclusion in the scientific community and remains so today -- the only way to resolve this dilemma was to create a new particle as being created in the whole event. Nevertheless, a very important question can be brought to mind: Q. Is there any guarantee that in a Neutron breakdown involving the Electron and Proton, that this event will exist as an even-steven scenario in the exchange of energy values? Perhaps the energy may just be the kinetic energy of the Electron speeding out of the environment rather than needing a completely new particle Neutrino?

<u>Kinetic energy</u> = the energy of an object due to its motion.

Note: The "missing mass" isn't really missing. It has been turned into energy.

In summary and looking at the above statements for each player involved in the Beta-decay event, the one main source for the development and potential distribution of any missing energy looks like it came from the **W-particle carrier force tasks of very high MASS (stated to be about 80,000 MeV)**.

QUARKS HISTORY

<u>Definition</u>:   quark (noun) – any of a number of subatomic particles carrying a fractional electric charge postulated as building blocks of the hadrons.   Quarks have not been directly observed, but theoretical predictions based on their existence have been confirmed experimentally.   Oxford Dictionaries

This will be a short history of quarks to simply ask: when, who, and why were quarks **invented**.   As I wrote earlier, I had no full comprehension of quarks when reading about them years ago, and I certainly do not intend to fully comprehend them at this particular time.   My only interest in this writing is to understand how they relate to a Beta-decay event and the Neutrino.

The names of all six quarks are: Up – Down – Strange – Charm – Beauty – Truth.   These names are known as quark "flavors."   Note: The word Beauty is now known as Bottom; the word Truth is now known as Top.   The first quarks of Up (U), Down (D), and Strange (S) were the announcement of Murray Gell-Mann and George Zweig, independently, in <u>1964</u>.   The name "quarks" is from Gell-Mann.   Further, quarks have also gone into dissection by referring to "colors" of blue, red, and green. This is all strictly for identification for the physicist.

Subsequently, the Charm (C) quark was the 4th type discovered in <u>1974</u>.   The Beauty or Bottom (B) was the 5th type discovered in <u>1977</u>.   The Truth or Top (T) was the 6th and last type discovered in <u>1995</u>, but it was a theory for 20 years.

In 1964, it was proposed that the Proton and Neutron particles could be better explained by having quarks "within" them; i.e., the Up quark and the Down quark. The values given to these two quarks were decided to be *fractional electric charges of 2/3 and -1/3*. Further it was decided that Protons and Neutrons will now be *made up of 3 quarks within each, in varied combinations of Up (U) and Down (D)*.

There is also something called "mesons" that contain quarks, two at a time, which need not be discussed here as they have no relationship to a Beta-decay event. Therefore, our focus will be on just a Proton (UUD) and a Neutron (UDD) containing only 3 quarks each.

It must be understood that even though these two huge particles contain 3 quarks each, the MASS energy of either the Proton and the Neutron is *supposedly not affected by having these quarks INSIDE THEM*. This is strictly what has been determined by the inventors of quarks. Nothing can be proven, because quarks have never been observed or measured or isolated on their own.

The last pieces of history regarding quarks can now be stated here. The inventor of quarks has determined that they are held together by something called "gluons" (which name implies that this would hold things together). Also, it has been determined that when one quark is pulled away from another quark, a "color force" will exist which involves the energy between the two quarks. These colors are stated as being blue, red, and green.

Note: This is where I must stop simply because this subject gets much more involved. It is tremendously arduous to follow.

Nevertheless, this whole scenario of quarks changing into other quarks only came about in <u>1964</u> when they were "invented" and "inserted" into the Neutron and the Proton community. Obviously, the Beta-decay event involving Protons and Neutrons had been going on long, long <u>before 1964</u>. However, it was simply referred to as a whole Neutron changed into a whole Proton. There was no mention of anything **inside** these huge particles to activate the transformation.

Last but not least, the whole picture of a Beta-decay event changed dramatically with the introduction of a carrier force particle which would help bring about this entire decay process. This particle would be known as a **Virtual W-particle.**

SOLVING THE NEUTRINO?

<u>Definition</u>: W particle (noun) – either of two particles about 80 times heavier than a proton that along with the Z particle are transmitters of the weak force and that can have a positive or negative charge [W+ or W-]. Merriam-Webster Dictionary

These particles were predicted as early as the late 1960s as having a large MASS which would interact with other particles and force a change. In a Beta-decay event, the heavy W-particle will perform in the breakdown of the Neutron as its

MASS/energy exerts itself over a tremendously short time. This function can never be observed. Its end product is the only proof of its accomplishment.

In 1984, Simon van der Meer, a Dutch engineer, along with Carlo Rubbia, an Italian physicist, received the Nobel Prize for Physics for the discovery of this massive, short-lived subatomic particle. By using the "electroweak theory" in this venture, they were able to estimate the masses of these W particles to be: nearly 80 to 100 times the MASS of the Proton!

(The Collins English Dictionary among its definition cites that the W particle has a rest mass of 1.435×10^{-25} kg.)

When a Neutron transforms into a Proton, it **emits** a W- particle. It is called a virtual particle because its MASS is so much greater than that of a Neutron and therefore the only way it can be emitted by this lighter particle Neutron is for its sheer existence to be drastically short-lived. A virtual particle is simply one that is **emitted and re-absorbed too rapidly to be detected**. It is ASSUMED BY REASON TO EXIST.

However, there is no known measurement system that can detect it! To be sure, it does not move very far from its emission point as it performs its tasks, and its work must be done rapidly because it will disintegrate practically immediately!

[Not wholly understood by the author: To say that this larger MASS Virtual W-particle was emitted by a lesser MASS particle Neutron is hard to comprehend. It somehow involves what is known as the "uncertainty principle."]

The explanation of the uncertainty principle is that this simply means that the more MASS and great energy of a particle, the greater will be the uncertainty of this energy and the shorter time permitted for its existence. That is to say, the principle of uncertainty clearly states that "position and momentum cannot be simultaneously determined with complete accuracy." In other words, the exact energy of an item (its momentum) cannot be determined at an exact time (its position). Needless to say, there will always be a "time interval" at which the "energy content" of an item is most definitely **uncertain**.

In a Beta-decay event, this larger MASS Virtual W-particle mediates or helps the Neutron, which just emitted it, in the immediate transforming of one Down quark into an Up quark which will produce a Proton and eliminate the Neutron. This is a big change.

But in doing this task, this Virtual W-particle will immediately transform **itself (or part of itself ?)** into an Electron of 0.000548, and, if believed, into a left-over energy for a Neutrino of 0.000291 (technically called an anti-neutrino in a Beta-decay event). Thus, the incredibly short life-span of the W-particle is done. It will disintegrate completely.

Two new theories with two different outcomes will be presented here: 1. Energy distribution was completed by Virtual W-particle absorption; or 2. Energy distribution was completed by a means other than creating a "ghost particle."

1. In plain English, it seems that there is **an explosion of the Neutron** in which one of the Down particles will not only transform into an Up particle, but this explosion has created more energy/MASS than what existed in the first place! Consequently, this <u>larger extra energy and larger MASS</u> of now the Virtual W-particle, which is surmised to spring forth from the Neutron Down particle, must quickly be put to tasks before this so-called "carrier force particle" will disappear/disintegrate.

The conversion of a Down particle into an Up particle is the primary function of the undertakings of this heavy-duty workhorse W-particle.   A secondary function is to transform some of its energy into a newly created Electron of a 0.000548 life force which will leave this environment and become in name a "Beta particle."   These happenings have taken place in an incredibly short period of time.   And, if believed, the creation of another particle (a so-called "ghost particle" Neutrino) of a minute 0.000291 life force will instantly take place from the large energy/MASS of the Virtual W-particle.   This named Neutrino was labeled for the left-over or "missing energy" in the decay process in which everything has to be accounted for by the tasks of the rapidly moving Virtual W-particle in order to satisfy the Conservation of Energy Law in its distribution scheme.

Nevertheless, before any solution can be applied, a most important question in all of the Beta-decay energy distribution has to be: Q. Where did any "missing energy" originate from? What is its source? The answer to that question is, of course, **the Virtual W-particle**. As stated previously as a definition, a <u>Virtual W-particle</u> is a subatomic particle with positive (W+) or negative (W-) charge that brings about the weak nuclear force.

Finally, a nagging question as to the distribution of this energy of 0.000291 must be asked:

Q. What if this so-called minute left-over energy of the totals of the Beta-decay event **could be placed into the action of a disintegrating W-particle**?

That is, when the W-particle deteriorates, it simply ceases to exist instantly, and the large **energy** that was in the MASS of this item is never able to be directly observed; only inferred. Therefore, it could be said that when this workhorse W-particle deteriorates after assisting in the creation of the Proton and in actually creating an Electron, it simply ceases to exist instantly wherein any **extra energy** that was in the MASS of this item is then **carried away by the W-particle in its disintegration**.

Could this be considered energy in the form of "kinetic energy" associated with the disintegration by the **power of the Virtual W-particle**? Any additional energy so calculated is simply absorbed upon disintegration?

In any case, all of the energy and MASS would be accounted for when the W-particle has ceased to exist in a Beta-decay event. There is nothing left of this carrier force Virtual W-particle. Most importantly, there is no hypothetical need for a new particle to be created at any time. There is no 0.000291 MASS/energy to be assigned to a new structure that has always been called a "ghost particle" because it cannot be detected by either electric charge or MASS. The bottom line is that the Neutrino can be eliminated as a creation in a Beta-decay event.

2. In plain English, the fact that an Electron having a negative electric charge is created by the Virtual W-particle in an instance and is not coming from the electrons residing around the nucleus, there is a distinct possibility that this newly created Electron could have a new MASS. That is to say, the newly created Electron would incorporate the 0.000291 energy into itself and be a completely new item. A new name will be suggested here for the purpose of identification in this writing: **Heavy Electron**.

This Heavy Electron was only needing a fleetingly new name as it was merely transformed/created/originating from the Virtual W-particle that performs its many tasks and disintegrates instantly. The Heavy Electron will be forever known as the Beta particle when it rapidly leaves the environment upon which it was created. No other particle will be created at this point.

It must be remembered that the size of the new Proton will be exactly that of any other Proton. However, a Beta particle originating through the Beta-decay event may not have any exactness hold over the creation of a normal sized electron. That is to say, this newly created Heavy Electron could most definitely **incorporate** any MASS/energy that the Virtual W-particle gives to it in its creation. This means that this new item could possess a life force of over and above that of an "Electron" MASS of 0.000548.

It could be theorized that this new Heavy Electron item has **absorbed all of the left-over energy (missing energy of 0.000291) after the creation of the new Proton by the Virtual W-particle**. Simplistically speaking, the sum total energy would now be locked into the new instant Heavy Electron creation. This new creation would have a value of: 0.000839. Interesting Note: The total of 0.000839 is very close to the sum total of an Electron (0.000548) and one-half an Electron (0.000274) for a total of: 0.000822.

Ultimately it must be remembered that this new Beta particle *will* interact with matter (unlike the fate of a Neutrino); it *will* travel at a decent rate of speed (unlike the massless fate of a Speed of Light Neutrino); and it *will not* go on forever in Space and Time (as is deemed the fate of a Neutrino particle).

Summary

As suggested herein, there are two options as to the extra energy/ MASS which has been calculated to occur in a Beta-decay event:

    1. the energy of 0.000291 **has been carried off** because it was simply absorbed by the Virtual W-particle upon its disintegration; or

    2. the energy of 0.000291 **was incorporated into a single new particle** having all of the left-over energy after the creation of a Proton by the Virtual W-particle. This fleetingly named particle will be flung out of its environment and continue on its way in the name of a "Beta particle."

Final Note: Particles are usually stated as having a purpose. That is to say, they co-exist in the world of science with some such inclusion into the scheme of things. But the most demeaning characteristic that has been given to this Neutrino particle is that it interacts with absolutely nothing! Consequently, if a Neutrino particle is to be treated as a duly created wave/particle that has no interaction whatsoever with anything valid to formulate with, then the conclusion must be drawn that it absolutely has no usefulness. Until this valueless particle can be shown as contributing somehow in the Universe in its mind-boggling output, estimated to be in the trillions upon trillions in numbers, then this Neutrino particle can be seen as having no purpose.

In my humble opinion, there is absolutely no logical reason and no function served for any present-day Neutrino to be characterized as a particle that is wildly roaming throughout the cosmos endlessly; i.e., with no absorption and with no useful interaction whatsoever. The Universe is built for interaction. Nothing escapes interaction.

Thus, the Neutrino needs to be examined with new thinking. ***

Respectfully, R.L. Dwyer

December 2017

SUGGESTED READING

Asimov, Isaac: *Understanding Physics, 3 Volumes in 1.* Dorset Press, 1966.

Asimov, Isaac: *The Neutrino (paperback).* Avon Books, 1966.

Berman, Bob: *The Sun's Heartbeat.* Little, Brown and Company, 2011.

Carson, Rachel L.: *The Sea Around Us.* Oxford University Press, Inc., 1950.

Clark, Stuart: *GALAXY, Exploring the Milky Way.* Fall River Press, 2008.

Cropper, William H.: *Great Physicists, The Life and Times of Leading Physicists from Galileo to Hawking.* Oxford University Press, 2001.

Dwyer, R.L.: *THE SUN CREATION THEORY presents the Creation of the Universe (contains 3 errors).* Strategic Book Publishing and Rights Co., 2014.

Feynman, Richard P.: *QED, The Strange Theory of Light and Matter.* Princeton University Press, 1985.

Grun, Bernard: *The Timetables of History, The New Third Revised Edition.* Simon & Schuster, 1946.

Newton, Isaac: *OPTICKS.* Great Mind Series, Prometheus Books, 2003.

Parker, Barry: *Einstein's Brainchild, Relativity Made Relatively Easy.* Prometheus Books, 2000.

Rees, Martin: *Before the Beginning.* Perseus Books, 1997.

* * *

INDEX

* * *